Mathematical Engineering

Today, the development of high-tech systems is unthinkable without mathematical modeling and analysis of system behavior. As such, many fields in the modern engineering sciences (e.g. control engineering, communications engineering, mechanical engineering, and robotics) call for sophisticated mathematical methods in order to solve the tasks at hand.

The series Mathematical Engineering presents new or heretofore little-known methods to support engineers in finding suitable answers to their questions, presenting those methods in such manner as to make them ideally comprehensible and applicable in practice.

Therefore, the primary focus is—without neglecting mathematical accuracy—on comprehensibility and real-world applicability.

To submit a proposal or request further information, please use the PDF Proposal Form or contact directly: Dr. Thomas Ditzinger (thomas.ditzinger@springer.com)

Indexed by SCOPUS, zbMATH, SCImago.

Alena Vagaská · Miroslav Gombár · Anton Panda

Optimization Methods in Mathematical Modeling of Technological Processes

Springer

Alena Vagaská
Department of Natural Sciences
and Humanities, Faculty of Manufacturing
Technologies
Technical University of Košice
Prešov, Slovakia

Miroslav Gombár
Department of Management, Faculty
of Management and Business
University of Prešov
Prešov, Slovakia

Anton Panda
Department of Automobile
and Manufacturing Technologies, Faculty
of Manufacturing Technologies
Technical University of Košice
Prešov, Slovakia

ISSN 2192-4732 ISSN 2192-4740 (electronic)
Mathematical Engineering
ISBN 978-3-031-35341-3 ISBN 978-3-031-35339-0 (eBook)
https://doi.org/10.1007/978-3-031-35339-0

This Springer imprint is published by the registered company Springer Nature Switzerland AG
The registered company address is: Gewerbestrasse 11, 6330 Cham, Switzerland

Acknowledgments

The authors thank the reviewers of the monograph for valuable advice and factual and formal comments that contributed to increasing the overall quality level of the publication. The authors also would like to thank the Scientific Grant Agency of the Ministry of Education, Science, Research and Sport of the Slovak Republic and Slovak Academy of Sciences for the support via the project VEGA 1/0226/21.

Introduction

The monograph focuses on the selected methods of applied mathematics aimed at mathematical optimization methods that have an important place in scientific research and in the professional profile of the twenty-first century engineers. Rapid development in the field of optimization methods and special software in connection with the growing computing power of modern computers has created and continues to create favourable conditions for the application of optimization in a wide range of scientific areas. Engineering optimization is of interest to many scientific research teams on a global scale. It is a part of the current mathematical modelling of processes and systems. Optimization modelling of technological processes is the focus of our attention. We discuss suitable linear, convex, and nonlinear optimization methods programming for solving problems of engineering practice. Implementation of methods is numerically illustrated on the solved problems applying real data using MATLAB software system and its extension to the convex optimization.

The optimization process is implemented using the selected algorithms (the Simplex method, IPM method, the Nelder–Mead method, and the BFGS). The aim of the monograph is to point out the possibilities of optimization in increasing the efficiency of the selected technological processes. Special attention is being paid to the optimization of cutting conditions during machining (turning) and optimization of surface treatment processes (galvanizing, anodic aluminium oxidation). The presented results give us the possibility to compare the models as well as to optimize the methods.

Aims of the Monograph

Having set the objectives, we focused on:

- Linking of mathematical optimization with computational software for the purpose of applications in the optimization of the selected technological processes. We are interested in:
- Optimization of cutting conditions during machining (longitudinal turning),
- Optimization of surface treatment processes (optimization of galvanizing process, minimizing the deposition time during anodic aluminium oxidation).

Contents

About the Authors

Alena Vagaská, doc. PaedDr. Ph.D. Professional graduation and growth: University of Pavol Jozef Šafárik in Košice, Slovakia (Mgr.—1993, mathematics and technical sciences); Constantine The Philosopher University in Nitra, Slovakia (PaedDr.—2003, Ph.D.—2004), DTI University, Slovakia (doc.—2021). She has 27 years of experience as a pedagogue and scientist at the Technical University of Košice, Faculty of manufacturing technologies with a seat in Prešov, Slovak Republic. She is the author and co-author of (269) publication outputs; specifically (9) monographs or chapters in monographs, (29) patents and discoveries, and (13) other book publications as university textbooks. She is also the author (co-author) of several foreign and domestic original scientific papers published in the scientific and professional journals (61), among them (14) are in the Impact Factored Current Contents Connect journals of WoS, in journals with ISI impact factors and publications included in the world-renowned databases—Thomson Scientific Master Journal List: Web of Science—(47), Scopus—(58). She focuses on research in applied mathematics, statistics, optimization, engineering manufacturing (and multidisciplinary), material sciences (multidisciplinary), and education/educational research. Research outputs from the above mentioned research fields have been presented at domestic and foreign scientific conferences and published as scientific and professional articles (152) in proceedings from conferences. She has been the head researcher (and co-researcher) in (26) domestic and international projects.

Miroslav Gombár, doc. Ing. Ph.D. Professional graduation and growth: Technical University of Košice, Slovakia (Ing.—2002, general engineering); Technical University of Košice, Faculty of Manufacturing Technologies, Slovakia (Ph.D.—2006, engineering technologies and materials), University of West Bohemia, Czech Republic (doc.—2018, mechanical engineering). He has 26 years of experience in surface treatment engineering practice (in a company aimed at surface treatment processes) and 15 years of experience in top management. For more than twenty years, he has been working as a teacher and scientist at several universities, let's mention the Technical University in Košice and the University of Prešov in Prešov. He is the author and co-author of (348) publication outputs; specifically (10) monographs or chapters in monographs, patents, and discoveries, and other book publications as university textbooks. He is also the author (co-author) of several foreign and domestic original scientific papers published in scientific and professional journals, among them, (27) are in the Impact Factored Current Contents Connect journals of WoS, in journals with ISI impact factors and publications included in the world-renowned databases—Thomson Scientific Master Journal List: Web of Science—(62), Scopus—(84). He focuses on research in engineering manufacturing (and multidisciplinary), material sciences (multidisciplinary), mechanical engineering, business and management, applied mathematics, statistics, optimization, and education/educational research. Research outputs from the above mentioned research fields have been presented at domestic and foreign scientific conferences and published as scientific and professional articles (182) in proceedings from conferences. He has been the head researcher (and co-researcher) in (27) domestic and international projects.

Anton Panda, Prof. Ing., Ph.D. University studies—Faculty of Mechanical Engineering TU Košice (Ing.—1987); terminated doctoral studies—Faculty of manufacturing technologies TU Košice (Ph.D.—2002), associate professor of study branch 5.2.51 manufacturing technologies, FMT TU Košice (assoc. prof.—2008), professor of study branch 5.2.51 manufacturing technologies, FMT TU Košice (prof.-2015). 29—years of experience in the engineering company supplying products for demanding automotive, also farm and agricultural industries (constructor of special purpose machinery and equipment, systems analyst, head of the department of development and technical preparation of production, methodist of statistical methods, commercial and technical director, director of quality). In the present expertise and design activities in the area of development, production, and verification of rolling bearings, in the area of deposition with rolling bearings for various domestic and foreign customers. Since 2008 (since 1994 external), operates as a pedagogue and scientist at the Faculty of manufacturing technologies TU Košice with a seat in Prešov, as well as an expert—coordinator (auditor) of quality management systems. He is the author (co-author) of 17 monographs (11 foreign, 6 domestic)—of this—3 monograph in Springer publishing, 2 university textbooks (1 foreign, 1 domestic), 16 university lecture notes, author's certificates (16), patents and discoveries (15), catalogs of bearings (2); several domestic and foreign original scientific papers in the scientific and professional journals, in Current Contents Connect journals in Web of Science (21), in impacted journals and publications led in the world renowned databases (Web of Science—117, Scopus—148) and in proceedings from domestic and foreign scientific conferences in the following research areas: automobile production, manufacturing technologies, experimental methods in the manufacturing technologies, machining, development, manufacturing, and verification of new products in accordance with the standards EN ISO 9001 and in accordance to the specific requirements of automobile manufacturers IATF 16 949, quality control, statistical methods and techniques of quality for the production of parts, capability of machine, capability of manufacturing processes, capability of gauges and measuring equipment, technical preparation of production, product audits, system audits of quality management system, analysis of potential errors and their effects on construction (FMEA-K) and on manufacturing process/technology (FMEA-V), statistical regulation of manufacturing processes SPC, process of approval of parts to the production PPAP, modern quality planning of product APQP, control plans and regulation, requirements the association of automobile manufacturers in Germany VDA 6.1, quality system requirements for suppliers of Ford, Chrysler, GM, specific requirements the using of EN ISO 9001:2015 in organizations ensuring the mass production in automotive industry IATF 16949, method of Poka-Yoke, quality assurance before the mass production for suppliers of automobile manufacturers in Germany VDA 4.3, quality assurance of supplies for suppliers of automobile manufacturers in Germany VDA 2, product liability, method of Global 8D (8-step method for solving of problems), etc. And these works are registered in the various domestic and foreign quotations and testimonials in the worldwide databases. Solver of several projects and grant projects for engineering companies at home and abroad, solver of research tasks, and author of the directives, methodological guidelines, technical regulations, and other technical documentation for domestic and foreign manufacturing companies. He is the auditor of quality system management on Technical University in Košice, Slovakia. Active collaboration with the university workplaces at home and abroad. He is recognized as an expert in the production of bearings in companies in Germany, Italy, China, Slovakia, and Czech Republic. As the coordinator of the research collective and co-author of documentation, EFQM has won the award for the improvement of performance in the competition National Award of Slovak Republic for quality in the year 2010 for the Technical University of Košice. In the same competition, he has won the same award in the year 2012, when the Technical University of Košice has obtained the highest score in its category. Since 2014 he has a member of the Polish Academy of Sciences. Since 2014, he has been the member of the ASME, USA.

Abbreviations

AAL	Layer Formed During AAO Process, also Denoted AAO Layer
AAO	Anodic Aluminium Oxidation
ACO	Ant Colony Optimization Method
BFGS	Quasi-Newton Method Developed by Mathematicians Broyden, Fletcher, Goldfarb, and Shanno
CCD	Central Composite Design
COST	Consider One Separate Factor at a Time
DFP	Davidon–Fletcher–Powell, the First Quasi-Newton Method
DOE	Design of Experiments
GA	Genetic Algorithm
GS	Method of Anodic Aluminium Oxidation in Sulphuric Acid at Direct Current
IPM	Interior Point Method
KKT	Karush–Kuhn–Tacker; KKT Optimality Conditions
lb	Lower Bound
LP	Linnear Programming
MP	Mathematical Programming
NP	Nonlinear Programming
OP	Optimization Problem
QNF	Quasi-Newton Formula
QNM	Quasi-Newton Method
QP	Quadratic Programming
SA	Simulated Annealing
SLP	Standard Linear Programming
ub	Upper Bound
VIF	Variance Inflation Factor

Chapter 1
Optimization in Historical Context

1.1 The Birth of Optimization as a Scientific Discipline

Optimization problems and tasks are widespread in virtually all areas of human activity and decision-making processes (examples can be found in economics, logistics, industrial production, medicine, nature itself, etc.). Such was the historical development and formation of optimization as a scientific discipline that is relatively young. Studying the history of this area, we can see that there are many independent lines of research and separately solved issues known as *optimal assignment, transportation problem, maximum flow, shortest spanning tree, shortest path, the traveling salesman problem,* etc.

It was not until the 1950s, when unifying tools for linear and integer programming became available and increased attention was paid to operational research and optimal management. These individual issues received a common framework and defined certain links and relationships that laid the foundations of the new scientific discipline.

But the roots of optimization go much deeper. We realize that the problems of optimal decision making are related to extreme value theory, i.e., the problems of setting a maximum and minimum value. The fact that there are several Latin words in this sentence (maximum—largest, minimum-smallest, extremum-extreme), suggests that the extreme value theory has been the subject of research since ancient times.

Foundation of the city of Carthage in 825 BC was associated with the aim to determine a closed curve that defines the area with the maximum possible surface in a plane. Similar problems have been referred to as isoperimetric. A characteristic feature of extreme problems is that their formulation was caused by urgent demands for the development of society [1].

A. Vagaská et al., *Optimization Methods in Mathematical Modeling of Technological Processes*, Mathematical Engineering, https://doi.org/10.1007/978-3-031-35339-0_1

1.2 From Traditional to Modern Optimization Methods

In any case, optimization is associated with an effort to minimize or maximize something that is simply expressed as an objective function. To determine an optimal solution of the objective function, many different optimization techniques such as linear programming at birth of optimization, simplex method, optimal assignment, transportation problem and many others that play an important role in the operation of companies, have gradually developed. Organizations implement these techniques to maximize profit, price from the point of view of the seller, quality, etc., or minimize cost, consumption, price from the buyer's point of view, duration of technological process, weight, etc.). We do not realize that in everyday life we solve various "optimization problems", for example when planning a holiday, we want to minimize costs and maximize the quality of services offered. In fact, permanently we are looking for optimal solutions to the problems we must face, even if we are not always able to find such the solutions. With any intention to optimize, it is necessary to realize that we can optimize only if we have possibility to choose from several variants. In the archaic approach, several suitable variants were recalculated, compared and consequently the only one variant was selected as the best. However, this process was time consuming and ultimately did not guarantee that the examined variants, which were considered appropriate, were not far from the optimal solution. That is why new optimization methods and algorithms have always been developed and tested.

To compare the effectiveness of optimization methods, different functions that caused various problems began to use. Among these there was the most famous *Rosenbrock's "banana function"*. Its modification, overview and comparison with the others can be found in [2–5]. In recent decades, rapid development of computer technology and its computing capabilities has made it possible to make more effective use of the existing modern optimization methods, techniques and models that currently enable us to effectively solve the problems in the order of millions of variables. The application field of modern optimization techniques such as engineering design, financial markets, fashion industry, mass communication, genetic engineering, is growing rapidly.

The complexity of the issue that is the subject of our optimization interest, makes it impossible to find any possible solution or combination of the solutions. The goal is to find the best acceptable solution from a feasible region to maintain certain conditions and restrictions. With some optimization techniques, there is no guarantee that the best solution will be found, and even we do not know if the algorithm will work properly and converge to the searched optimal solution. Difficulties arise in optimization of large-scale problems and problems caused by multi-modality, dimensionality, and differentiability. In general, traditional techniques fail to solve complex and extensive problems, especially in connection with nonlinearity of the objective function. Serious problem with the use of traditional techniques is also the indistinguishability of the objective function since most traditional optimization techniques require knowledge of the gradient that is impossible with this type of the objective function. In addition, such techniques appear unreliable even in case of existence of several local optima. To tackle these issues, we need to develop stronger and more reliable, more *robust optimization techniques* that are known also under the concept of *modern optimization techniques.*

They are used to solve linear, nonlinear, differentiable, and non-differentiable optimization problems. The above facts initiated the development of *combinatorial* or *statistical optimization* as well as metaheuristic methods.

A. Turning was probably the first to use a heuristic algorithm. During World War II he deciphered the German code Enigma. In 1940 in Bletchley Park, together with the British mathematician G. Welchman, they designed cryptanalytic electromechanical machine called the Bombe. They used a heuristic algorithm, as Turning called it, searching between 1022 potential combinations the correct code setting in the Enigma message. Heuristic or discovery process is a process that is based not only on logical reasoning and experience, but also on observation and experiments. Turning named his method heuristic search method because finding a solution can be expected. The method works very successfully in most cases even if there is no guarantee and evidence of finding the right solution [6].

In addition to heuristic methods, the 1960s were also successful for *interior point methods* (IPM), which have been widely cultivated and analysed particularly in the 1960s under nonlinear programming, but they were unsuccessfully applied to this type of problems and then forgotten. However, a reversal in the fate of these methods occurred in 1984, when Karmakar published his famous algorithm for solving linear programming problems, thereby renewing interest in interior point methods leading to the so-called *optimization revolution.* He influenced the development not only in linear programming, but also the whole mathematical programming and optimization. The year of Karmakarov's publication of his algorithm (1984) began the *era of modern interior point methods* in optimization. The idea of interior point methods is fundamental to algorithms, often referred to as *path-following algorithms* [7].

Another important step was the development of evolutionary algorithms in 60s and the 70s. In 1962, J. Holand studied adaptive systems and together with his colleagues from Michigan university developed the *genetic algorithm* (GA). He was the first to use crossover and recombination manipulations to model such systems. The philosophy of GA is based on Darwin's theory of evolution (theory of survival of the fittest). Genetic algorithms are based on the principles of genetics and natural selection. The basic elements of reproduction—crossing and mutation are used in the GA procedure.

Various optimization techniques are being developed to tackle problems and obstacles we face in industry or everyday life. In case of various mixed problems, development of *metaheuristic algorithms* was mostly inspired by nature based on the abstraction of its functioning. For these techniques there always exists the most appropriate solution. Nature has evolved millions of years and always found the perfect solution to almost any problem. As a result of these facts, such algorithms refer to as *biological algorithms* or bio algorithms [6].

The two basic components of each metaheuristic algorithm are: *selection of the best* solution and *randomization.* Choosing the best guarantees that solutions lead to the optimum, while randomization ignores locally optimal solutions and increases the diversity of solutions. Effective use of these two components usually ensure that a global optimum is achievable.

The decades of 1980 and 1990 were especially graceful and exciting for metaheuristic algorithms. Another important milestone in history of optimization was the development of a *simulated annealing* (SA) method as an optimization technique (1983). Its pioneers were S. Kirkpatrick, C. D. Gellat and M. P. Vecchi who were inspired by the process of steel annealing [8]. This method belongs to the so-called *path-following algorithms*. They require an initial solution estimate at start-up and the so-called initial point (high temperature) must be set up. Gradually, the system cools down. In the next step of the algorithm, we want to approach the optimum, so a movement or a new solution is acceptable if it is better. However, it is acceptable with some probability that causes the system to skip some local optima. We suppose that if the system is cooled sufficiently slowly, a global optimal solution will be achieved [9].

In 1992, M. Dorigo completed his PhD thesis where he presented its innovative approaches to *ant colony optimization* (ACO). This search technique was inspired by a group intelligence of the ant society using pheromones as a chemical message [10].

Let us mention some other modern optimization techniques, such as *Honey Bee Algorithm, Greedy Algorithm, Tabu Method, colony intelligence algorithms, artificial intelligence algorithms*, algorithms based on the use of Fuzzy sets and logic and other nature and biology inspired algorithms such as *ecological algorithms* or *biological algorithms* [6, 10–13].

In the next chapter of the monograph, we will present basic information on optimization problems, definition of terms, formulation of optimization problems and their classification. In the following chapters we will focus on the selected types of optimization tasks in connection with the solution of technological processes optimization.

We will gradually discuss the types of problems pointing out the possibilities of applying selected optimization methods and techniques using MATLAB software to solve engineering issues.

References

1. Колманоьский, В. Б. (1997). *Задачи оптимального управления. Математика*. [Kolmanovskii, V. B. *Problems of optimal control. mathematics*] (pp. 121–127). Московский государственный институт електроники и математики, Москва.
2. Cao, H., Qian, X., & Zhou, Y. (2018). Large-scale structural optimization using metaheuristic algorithms with elitism and a filter strategy. *Structural and Multidisciplinary Optimization, 57*, 799–814 (2018). https://doi.org/10.1007/s00158-017-1784-3.
3. Hudzovič, P. (2001). *Optimalizácia* (320p.). STU v Bratislave, Bratislava, Slovakia. ISBN 80–227–1598–0.
4. Реклейтис, Г., Рейвиндран, А., & Рэгсдел, К. (1986). *Оптимизация в технике* 1, [Reklaitis, G. V., Ravindran, A. Ragsdell, K. M. *Engineering optimization. Methods and applications*] (350p.). Mir, Moskva.

5. Реклейтис, Г., Рейвиндран, А., & Рэгсдел, К. (1986). *Оптимизация в технике* 2, [Reklaitis, G. V., Ravindran, & A. Ragsdell, K. M. *Engineering optimization. Methods and applications*] (320p.). Mir, Moskva.
6. Khan, S., Asjad, M., & Ahmad, A. (2015). Review of modern optimization techniques. *International Journal of Engineering Research & Technology (IJERT), 4*(04). https://doi.org/10.17577/IJERTV4IS041129.
7. Halická, M. (2004). Dvadsať rokov moderných metód vnútorného bodu. *Pokroky matematiky, fyziky a astronomie, 49*(3), 234–244.
8. Kirkpatrick, S., Gelatt, C. D., & Vecchi, M. P. (1983). Optimization by simulated annealing. *Science, 220*(4598), 671–680. https://doi.org/10.1126/science.220.4598.671
9. Černý, V. (1985). Thermodynamical approach to the traveling salesman problem: An efficient simulation algorithm. *Journal of Optimization Theory and Applications, 45*(1), 41–51.
10. Dorigo, M. (1992). *Optimization, learning and natural algorithms*. Ph.D. thesis. Politencnicodi Milano, Italy.
11. Dybvig, P. H. (2012). Numerical methods for optimization. In *Fin500J mathematical foundations in finance* (25pp).
12. Yang, Z. (2021). On the step size selection in variance-reduced algorithm for nonconvex optimization. *Expert Systems with Applications, 169* (2021), 114336, 12p. https://doi.org/10.1016/j.eswa.2020.114336.
13. Zhou, D., Xu, P., & Gu, Q. (2018). Stochastic nested variance reduction for nonconvex optimization. In *International conference on neural information processing systems* (pp. 3925–3936). Curran Associates Inc.

Chapter 2
Optimization

2.1 Mathematical Foundations of Optimization

Optimization can be defined in various ways. Simplified perception of optimization as "the art of doing things best" [1, 2] is presented by many critics who say that in the real world it is not possible to do something perfectly. There are always some restrictions. In practice, however, it is desirable to do something *as good as possible within practical constraints*. In the broadest sense, optimization means *getting the best results under certain conditions*, it represents a selection effort to find the best variant of solving the problem to find the optimal variant [3]. Within such an understanding, optimization is an integral part of every human being activity carried out either intuitively or with a greater degree of awareness (for example, minimizing shipping time and costs is a matter of course for each) [2, 4, 5].

Without the use of optimization tools and methods, this approach is *intuitive*. The best results can also be achieved with experience background, intuition and with a certain amount of happiness [2]. Optimization can be considered as a *systematic process of finding the optimal variant* where tools and methods of mathematics and the power of computer technology are used. Today, there exists relatively good and widely elaborated theory of optimization methods and procedures supported by software because of their algorithmic property [3]. Then the focus of the optimization problem moves to the area of its formulation. Correct formulation of optimization must not be underestimated as it affects success or failure. Even with the exact solution of an incorrectly formulated problem, the obtained solution will not be optimal [3, 4].

If for optimization the exact meaning of the term is defined, which is expressed *mathematically*, optimization acquires a solid foundation for a *scientific approach* to solving technological, control, organizational and decision-making problems [4]. In essence, control or organization is the implementation of a decision, and this must be such that of all possible solutions it coincides with the best solution, i.e., the optimal solution [3]. Therefore, optimization is only possible if we have the option to choose

A. Vagaská et al., *Optimization Methods in Mathematical Modeling of Technological Processes*, Mathematical Engineering, https://doi.org/10.1007/978-3-031-35339-0_2

a solution of several variants which are generally not equivalent. Regarding with the above we are talking about *optimal decision-making*, i.e., *optimal control.*

The task of process optimization is to find the best solution using a suitable optimization method [2, 6–10].

The object of optimization is to provide decision-making tools on choosing the best possible solution to specific tasks that may be of various character [5, 11–13].

From a mathematical point of view, optimization is a process involved by finding such points of the function in which the function reaches the extreme (maximum, minimum). To master the issue of finding extremes as well as to understand the following chapters it is necessary to introduce several definitions which follow from the literature [14–21]. Marking of *vector variables* in bold is currently being abandoned and for clarity of relationships and without prejudice to validity and generality we will consider it in later chapters.

2.1.1 Minimum of a Function

Definition 2.1.1 Let $f : X \to \mathbb{R}$ be a defined function. Point $\boldsymbol{x}^* \in \mathbb{R}^n$ is called

(i) a point of a **local minimum** of a function f on $X \subset \mathbb{R}^n$ if there exists a number $\delta > 0$ so that

$$f\left(\boldsymbol{x}^*\right) \leq f(\boldsymbol{x}) \quad \forall \boldsymbol{x} \in X,\ \left\|\boldsymbol{x} - \boldsymbol{x}^*\right\| < \delta$$

(ii) a point of a sharp **local minimum** of a function f on $X \subset \mathbb{R}^n$ if there exists a number $\delta > 0$ so that

$$f\left(\boldsymbol{x}^*\right) < f(\boldsymbol{x}) \quad \forall \boldsymbol{x} \in X, 0 < \left\|\boldsymbol{x} - \boldsymbol{x}^*\right\| < \delta$$

(iii) a point of a **global minimum** of a function f on $X \subset \mathbb{R}^n$ if

$$f\left(\boldsymbol{x}^*\right) < f(\boldsymbol{x}) \quad \forall \boldsymbol{x} \in X$$

(iv) a point of a **sharp global minimum** of a function f on $X \subset \mathbb{R}^n$ if

$$f\left(\boldsymbol{x}^*\right) < f(\boldsymbol{x}) \quad \forall \boldsymbol{x} \in X,\ \boldsymbol{x} \neq \boldsymbol{x}^*$$

Remark 2.1.1 If in *(i)–(iv)* opposite inequalities are valid for $f(\boldsymbol{x}^*)$ and $f(\boldsymbol{x})$, we say that at a point $\boldsymbol{x}^*$ a function f acquires *(i)* ***local maximum***, *(ii)* ***sharp local maximum***, *(iii)* ***global maximum*** and *(iv)* ***sharp global maximum***. The condition $0 < \|\boldsymbol{x} - \boldsymbol{x}^*\| < \delta$ expresses δ-surrounding of the point $\boldsymbol{x}^*$, which is an open sphere with radius δ and centre at $\boldsymbol{x}^*$, is also written in the form $\mathcal{O}_\delta(\boldsymbol{x}^*)$.

2.1.2 Gradient and Hessian

Definition 2.1.2a Let $f \in C^1$. By a **gradient** of a function f at a point $\boldsymbol{x} \in \mathbb{R}^n$ we understand a vector

$$\nabla f(\boldsymbol{x}) = g(\boldsymbol{x}) = \begin{pmatrix} \frac{\partial f}{\partial x_1}(\boldsymbol{x}) \\ \vdots \\ \frac{\partial f}{\partial x_n}(\boldsymbol{x}) \end{pmatrix}$$

Definition 2.1.2b Let $f \in C^2$. By the **Hessian** of a function f at a point $\boldsymbol{x} \in \mathbb{R}^n$ we understand symmetric matrix

$$\boldsymbol{H}(\boldsymbol{x}) = \left(h_{ij}(\boldsymbol{x})\right)_{i,j=1}^{n} \quad h_{ij}(\boldsymbol{x}) = \frac{\partial^2 f}{\partial x_i \partial x_j}(\boldsymbol{x})$$

$$\boldsymbol{H} = \begin{pmatrix} \frac{\partial^2 f}{\partial^2 x_1} & \frac{\partial^2 f}{\partial x_1 \partial x_2} & \cdots & \frac{\partial^2 f}{\partial x_1 \partial x_n} \\ \frac{\partial^2 f}{\partial x_2 \partial x_1} & \frac{\partial^2 f}{\partial^2 x_2} & \cdots & \frac{\partial^2 f}{\partial x_2 \partial x_n} \\ \vdots & \vdots & \ddots & \vdots \\ \frac{\partial^2 f}{\partial x_n \partial x_1} & \frac{\partial^2 f}{\partial x_n \partial x_2} & \cdots & \frac{\partial^2 f}{\partial^2 x_n} \end{pmatrix}$$

Remark 2.1.2 We use the key properties of the gradient and the Hessian in optimization. Gradient $\nabla f(\boldsymbol{x})$ determines the descent of a function f at a point $\boldsymbol{x}$, by $\nabla f(\boldsymbol{x})$ we determine the direction of the fastest ascent of a function, by $-\nabla f(\boldsymbol{x})$ we determine the direction of the steepest descent of a function. The Hessian $\boldsymbol{H}(\boldsymbol{x})$ determines the shape of a function f around a point $\boldsymbol{x}$.

2.1.3 Definiteness and Semidefiniteness of a Matrix

A matrix $\boldsymbol{A} = \left(a_{ij}\right) \in \mathbb{R}^{n \times n}$ is symmetric if for $\forall i, j$ the following is valid: $a_{i,j} = a_{j,i}$.

Definition 2.1.3a Real symmetric matrix $\boldsymbol{A}$ is called **positive definite** if for any complex or real n-dimensional vector $\boldsymbol{x}$ the following is valid:

$$\boldsymbol{x}^T \boldsymbol{A} \boldsymbol{x} > 0 \quad \boldsymbol{x} \neq 0$$

Definition 2.1.3b Real symmetric matrix $\boldsymbol{A}$ is called **positive semidefinite** if for any complex or real n-dimensional vector $\boldsymbol{x}$ the following is valid:

$$\boldsymbol{x}^T \boldsymbol{A} \boldsymbol{x} \geq 0 \quad \boldsymbol{x} \neq 0$$

Remark 2.1.3 The definiteness of a quadratic form $x^T Ax$ is assessed analogously, where A is a symmetric matrix. If matrix A is positive (semi)definite, then $B = -A$ is negative (semi)definite. To determine a positive definite directly from the definition is sometimes problematic, so eigenvalues λ_i of a matrix $\mathbf{A}$ are used, or it is verified whether the Sylvester's criterion apply to all minors of a matrix [11]. We will introduce later that positive (negative) semidefiniteness of the Hessian matrix $H(x)$ relates to convexity (concavity) of an investigated function f.

2.1.4 Stationary Point

Definition 2.1.4 Let $f \in C^1$. Point $\widehat{x} \in \mathbb{R}^n$ is a stationary point of a function f if the following is valid

$$\nabla f\left(\widehat{x}\right) = g\left(\widehat{x}\right) = 0$$

Remark 2.1.4 A gradient is a column vector of the first order partial derivatives according to all variables. Then a function can take on an extreme only at those points where it has all partial derivatives of the first order equal to zero (at stationary points) or at points where it has no derivative.

2.1.5 Convexity

Definition 2.1.5a A set $M \subseteq \mathbb{R}^n$ is called convex if for each of the two points $x, y \in M$ the following is valid

$$\alpha x + (1 - \alpha)y \in M \quad \forall \lambda : \quad 0 \leq \alpha \leq 1$$

Definition 2.1.5b A function $f(x)$ defined on a convex set $M \subseteq \mathbb{R}^n$ is convex if for any $x, y \in M$ the following is valid

$$f(\alpha x + (1 - \alpha)y) \leq \alpha f(x) + (1 - \alpha)f(y) \quad \forall \alpha : \quad 0 \leq \alpha \leq 1$$

In addition, if a sharp inequality applies to $\forall \alpha \in (0, 1)$ and $\forall x, y \in M$, $x \neq y$, then we say that f is strictly convex to M.

Theorem 2.1.5 *Let $f \in C^2$, $M \subseteq \mathbb{R}^n$ be a convex set with nonempty interior. Function f is convex on M if only its Hessian is positive semidefinite on M. Function f is strictly convex on M if only its Hessian is positive definite on M.*

Remark 2.1.5 In optimization problems, convexity property of a set and convexity or concavity of a function is very important. When solving optimization problem, where we must find the optimum (minimum) of a convex function, the respective optimization procedures are significantly less computationally demanding if it is a nonconvex problem. Many optimization methods follow from the convexity of a problem, although it is not always directly stated [5, 14, 22, 23].

2.1.6 Descent Direction

Definition 2.1.6a Let $f(\boldsymbol{x}) : \mathbb{R}^n \to \mathbb{R}$ be a given function, $\boldsymbol{x} \in \mathbb{R}^n$ a given point and vector $\boldsymbol{p} \in \mathbb{R}^n$ a given direction. If there exists $\alpha > 0$ such that

$$f(\boldsymbol{x} + \alpha \boldsymbol{p}) < f(\boldsymbol{x})$$

then $\boldsymbol{p}$ is called a descent direction of function f at a point $\boldsymbol{x}$.

Definition 2.1.6b Let $f \in C^1$ be a given function, $\boldsymbol{p} \in \mathbb{R}^n$ is a given direction and let $\boldsymbol{g}(\boldsymbol{x})$ denote a gradient of a function f at a point $\boldsymbol{x}$. If the following is valid

$$\boldsymbol{g}(\boldsymbol{x})^T \boldsymbol{p} < 0$$

then $\boldsymbol{p}$ is a descent direction of f at a point $\boldsymbol{x}$.

2.1.7 Convergence Rate

Definition 2.1.7 Let a sequence $\{\boldsymbol{x}_k\}$ converges to a point $\boldsymbol{x}^*$. This applies to a convergence rate

$$\lim_{k \to \infty} \frac{\|\boldsymbol{x}_{k+1} - \boldsymbol{x}^*\|}{\|\boldsymbol{x}_k - \boldsymbol{x}^*\|^p} = c$$

where $0 < c < \infty$ is called an error constant. Convergence rate is linear for $p = 1$ and $c < 1$, for $p = 2$ quadratic and for $p = 3$ it is cubic. If $p = 1,\ c = 0$ or $1 < p < 2,\ \ c > 0$ it is a superlinear convergence.

2.2 Formulation of an Optimization Problem

Mathematical formulation of an optimization problem is generally based on the search for the minimum (when changing the sign, then we search for the maximum) of a function which is known as an objective function or criterion function, target function, optimality criterion or optimization functional [2–5, 10, 14, 24, 25]. We most often encounter the presented formulations in the literature.

A problem of mathematical programming (MP) or just an **optimization problem** (OP) generally means finding a solution to a problem

$$\min\{f_0(\boldsymbol{x})|\boldsymbol{x} \in X \}, kde f_0 : X \to \mathbb{R} \ aX \subseteq \mathbb{R}^n, \tag{2.1}$$

This is an optimization problem of *minimization.*

A vector $\boldsymbol{x} = (x_1, \ldots, x_n)^T$ is an optimization variable, i.e. a *design vector*.

A function $f_0 : X \to \mathbb{R}$ is called an *objective function*, X is a *feasible region* and each of its elements $\boldsymbol{x} \in X$ is called a *feasible solution.*

If a set X is empty, i.e. $X = \emptyset$, we say that the optimization problem is *infeasible.*

A vector $\boldsymbol{x}^* \in X$ is called *optimum* or *optimal solution* to a problem (2.1) if it assumes the smallest value among all the vectors in the feasible region.

If there is a sequence $\boldsymbol{x}^k \in X$ so that $f\left(\boldsymbol{x}^k\right) \to \infty$ for $k \to \infty$, we are talking about an unconstrained problem. In this monograph we deal with the so-called finite dimensional problems for which $X \subset \mathbb{R}^n$.

Depending on a feasible region X, we speak of *unconditional* and *conditional* optimization problems.

If $X = \mathbb{R}^n$ in the optimization problem (2.1), then any point $\boldsymbol{x} = (x_1, \ldots, x_n)^T$ of n-dimensional Euclidean space $\mathbb{R}^n$ may be a feasible solution. In this case we speak of *unconditional optimization.* If X is a custom subset of space $\mathbb{R}^n$, i.e., $X \subset \mathbb{R}^n$, we speak of *conditional optimization.* In this case, the set X expresses a *feasible region of the solution* through conditions specified in the form of equations and inequalities.

In many publications, we can find deviations in the definition of mathematical programming problems, but the essence remains the same—an extremization of a function. For many authors, the differences are mainly in the marking and the symbols used. The authors [23] use a symbol *Min* (with a capital initial letter) instead of the usual operator *min* expressing a final state to emphasize that a problem of optimization (minimization) is the process of finding a final state. Denoting of vector variables has been updated (currently denoting vectors using boldface is omitted). To unify the symbolism in the following text of the monograph, we follow the definitions presented in [14, 23].

Definition 2.2.1 The task to **minimize an objective function** $f_0(\boldsymbol{x})$ in a feasible region K is formulated as **a mathematical programming problem** and is written as follows

$$Min\left\{f_0(\boldsymbol{x})\middle|\boldsymbol{x} \in K \subseteq \mathbb{R}^n\right\} f_0 : X_0 \subseteq \mathbb{R}^n \to \mathbb{R} \tag{2.2}$$

if feasible region K has the form

$$K = \{\boldsymbol{x} \in X\,|\,f_i(\boldsymbol{x}) \leq 0, i = 1, 2, \ldots, m\} \neq 0 \tag{2.3}$$

where

$$f_i : X_i \to \mathbb{R}, i = 0, 1, 2, \ldots, m, \emptyset \neq X_i \subseteq \mathbb{R}^n \text{ a } X = \bigcap_{i=0}^{m} X_i$$

We can briefly write a **mathematical programming problem** (MP) in the following form

$$Min\{f_0(\boldsymbol{x}) | \boldsymbol{x} \in X, f_i(\boldsymbol{x}) \leq 0, i = 1, 2, \ldots, m\} \tag{2.4}$$

and we call it the MP problem in the narrower sense.

If $\boldsymbol{x} \in X$ where $X \subset \mathbb{R}^n$ we are talking about conditional optimization.

For $\boldsymbol{x} \in \mathbb{R}^n$ this is a problem of unconditional optimization.

From the Definition 2.2.1 it is clear that X is the intersection of all $m + 1$ domains X_i of functions f_i which is assumed to be a nonempty subset of space $\mathbb{R}^n$, i.e. $\emptyset \neq X \subseteq \mathbb{R}^n$, where X_i are open space sets and there is no relation between a number of m constraints and a number of n variables. Then it is obvious that $\boldsymbol{x} = (x_1, x_2, \ldots, x_n)^T$ is a vector of independent optimization variables and a feasible region K is given by a finite number of inequalities and equations considered on a certain set $X \subset \mathbb{R}^n$.

An inequality $f_i(\boldsymbol{x}) \leq 0$ is called the *i-constraint of the problem* (2.4).

Optimal solution to a problem (2.4) is a such feasible solution $\boldsymbol{x}^* \in K$ for which the following is valid

$$\forall \boldsymbol{x} \in K : f_0\left(\boldsymbol{x}^*\right) \leq f_0(\boldsymbol{x}) \tag{2.5}$$

If we substitute an optimal solution $\boldsymbol{x}^* \in K$ into the individual boundaries, then constraints that are realized as equalities $f_i(\boldsymbol{x}^*) = 0$ are called active constraints (considering an optimal solution $\boldsymbol{x}^* \in K$). Other constraints that are realized as sharp inequalities $f_i(\boldsymbol{x}) < 0$ are called *inactive constraints* [26]. Constraint conditions expressed by functions $f_i(\boldsymbol{x})$ are called *functional constraints* (express the conditions and constraints under which the process under consideration is to operate), condition $\boldsymbol{x} \in X$ is called a direct constraint [27] if the problem $X = \mathbb{R}^n$ has no direct constraints.

Remark 2.2.1 In the literature, we often encounter the definition of a mathematical programming problem in such a form that in addition to m constraints in the form of the following inequalities

$$f_i(\boldsymbol{x}) \leq 0, \ i \in I = \{1, 2, \ldots, m\}$$

r-constraints in the form of the following equations are also considered

$$h_j(\boldsymbol{x}) = 0, \ j \in J = \{m + 1, m + 2, \ldots, m + r\}$$

where I and J are index sets.

However, this complicates the symbolism, and the derived relationships are less clear. The solution to how to avoid this fact is proposed by the authors [23] who prefer the above procedure and for clarification purpose they call the problem (2.4) a *mathematical programming problem (MP) in the narrower sense* and an analogous problem containing constraints in the form of the equations

$$Min\{f_0(\boldsymbol{x}) \;|\boldsymbol{x} \in X, f_i(\boldsymbol{x}) \leq 0, i \in I,\; h_j(\boldsymbol{x}) = 0, j \in J\} \tag{2.6}$$

is called a *mathematical programming problem (MP) in the broader sense.* Sets I, J are index sets in the form $I = \{1, 2, \ldots, m\}$ and $J = \{m+1, m+2, \ldots, m+r\}$.

Definition 2.2.2 Function minimization $f_0(\boldsymbol{x})$ is understood as a process of searching for some minimum of this function.

Instead of a *minimization* problem, it is possible to consider a *maximization problem.*The validity of the relationship

$$\min_{x\in\mathbb{R}^n} f_0(\boldsymbol{x}) = -\max_{x\in\mathbb{R}^n}(-f_0(\boldsymbol{x})) \tag{2.7}$$

allows us to transform each maximization problem into a minimization problem. That is why in the next part of the monograph we can limit ourselves only to minimization. From the above it is clear that it does not matter what type of extremalization we consider to be a standard. From the Definition 2.1.1 and the Remark 2.1.1 we know that these are global or local minima.

2.3 Classification of Optimization Problems

When solving optimization problems in engineering area, it is important to determine the category of an optimization problem. This fact subsequently influences the selection of a suitable optimization method and thus the efficiency of the whole optimization process as there is no single universal method for solving various optimization problems. Optimization problems can be characterized and divided into individual classes or types according to various aspects [11, 14, 20, 23, 25, 27–33].

Classification of problems according to the nature of a feasible region (existence of constraints)

Unconstrained optimization problems

This situation occurs, when $K \subset \mathbb{R}^n$ in (2.2), i.e. feasible region K is an open set. According to the Definition 2.2.1 this is an *unconditional optimization.*

Constrained optimization problems

When $K \subset \mathbb{R}^n$ in (2.2), i.e. feasible region K is a closed set determined explicitly by a system of real equations and inequalities, then occurs constrained OP expressed by (2.6). According to the Definition 2.2.1, this is a *conditional optimization,* and individual variables must meet the prescribed limits which define a feasible region.

Classification according to the type of functions (special problems of mathematical programming)

Linear programming—occurs when all the functions $f_0,\ f_i,\ i \in I,\ h_j, j \in J$ are linear, i.e. if in optimization problem (2.6) all $m + r$ functions are linear (both the objective function f_0 and all the functions of the constraints f_i and h_j).

Nonlinear programming—occurs when at least one of the functions $f_0,\ f_i,\ i \in I,\ h_j,\ j \in J$ is not linear.

Convex programming—when functions $f_0,\ f_i,\ i \in I$ in (2.6) are convex and functions $h_j,\ j \in J$ are linear.

The following applies to mathematical programming problems

$$LINEAR \subseteq CONVEX \subseteq NONLINEAR$$

Quadratic programming is a nonlinear programming which occurs when functions of the constraints $f_i,\ i \in I,\ h_j,\ j \in J$ are linear and the objective function f_0 is quadratic with a positive semidefinite symmetric matrix $\boldsymbol{Q}$.

Geometric programming—works with the so-called pozinomas of the shape $\sum_{i=1}^{m} c_i \prod_{j=1}^{n} x_j^{a_{ij}}$, where $\boldsymbol{x} \in \mathbb{R}_+^n$ and $\boldsymbol{c} \in \mathbb{R}_+^m$ are vectors whose components are positive.

Discrete programming—works with a discrete feasible region K.

Integer programming—works with an integer set X.

Partly integer programming—combines integer variables with continuous ones.

Stochastic programming includes problems involving uncertainty.

Continuous programming—works with continuous variables.

Let us present a few facts about individual classes or types of optimization problems which are characterized by a certain form of an objective function and constraint functions. We stated above that the optimization problem called *linear programming* includes an objective function and constraint functions which are linear. Function f_i is called linear if it satisfies

$$f_i(\alpha \boldsymbol{x} + \beta \boldsymbol{y}) = \alpha f_i(\boldsymbol{x}) + \beta f_i(\boldsymbol{y}) \tag{2.8}$$

for all $\boldsymbol{x},\ \boldsymbol{y} \in \mathbb{R}^n$ and all $\alpha,\ \beta \in \mathbb{R}$

For *convex optimization* problems, an objective function and constraint functions of problem (2.6) are convex in the sense of the Definition 2.1.5b and, in addition, they satisfy the inequality

$$f_i(\alpha \boldsymbol{x} + \beta \boldsymbol{y}) \le \alpha f_i(\boldsymbol{x}) + \beta f_i(\boldsymbol{y}) \tag{2.9}$$

for all $\boldsymbol{x},\ \boldsymbol{y} \in \mathbb{R}^n$ and all $\alpha,\ \beta \in \mathbb{R}$ where $\alpha + \beta = 1,\ \alpha \ge 0,\ \beta \ge 0$.

By comparing (2.8) and (2.9) we can see that convexity is more general than linearity: inequality replaces much stricter equality, and inequality must apply only to certain values of α and β. Therefore, any linear programming is a convex optimization problem. Convex optimization can be considered as a generalization of linear programming [14]. Criterion of dividing optimization problems into *convex* and *nonconvex problems* is important, as

- *convex optimization problems* are computationally simpler as formulation of a problem guarantees a single extremum, i.e., a local extremum is also a global extremum;
- *nonconvex optimization problems* have a nonconvex objective function, or feasible region is nonconvex. When solving nonconvex problems, approximation of nonconvex area by convex is often used to apply the methods of solving convex problems.

Classification according to the nature of variables (optimization variables, objective function)

Continuous optimization problems—in the optimization problem (2.4), the objective function is continuous, i.e., continuous values of optimization parameters are allowed. Only isolated points of discontinuity are allowed.

Discrete optimization problems—variables can only take discrete values (integer programming, binary problems with variables values 0 and 1).

In general, it is a characteristic feature of discrete problems that variables are selected from a finite, even though arbitrarily a large set or from a countable set. In contrast, continuous problems are characterized by variables selected from an infinite uncountable set (typically from a set of real numbers). Then, the way of solving these tasks is related to this fact. When solving continuous problems, the advantage lays is the possibility to use continuity or smoothness of the relevant functions. Considering information about the objective function and the constraints at a particular point, the behaviour of these functions in the "immediate vicinity" can be assumed. In contrast to this typical property of continuous problems, with discrete problems the behaviour of the objective function and the constraints can change significantly when moving from one point of a feasible region to another point. This significantly complicates the solution of an optimization problem.

Classification according to the nature of a mathematical model of a system

Static optimization problems—steady states without time (developmental) dependencies are modelled and a *static mathematical model* of a problem is created. The aim is to achieve optimal values of quantities in a steady state which will ensure the best technological and economic result of control. A range of methods have been developed to solve the problems of this type including mathematical programming problems (nonlinear MP, linear MP).

Dynamic optimization problems—solve optimization problems for a dynamic system where time dependences of quantities are also modelled, and a *dynamic model* of the situation is created. The aim is the optimal course of transients of a controlled variable when changing control or fault quantities. This includes the issue of variational calculus (the Pontrjagin's principle, dynamic programming). With this classification, we consider the so-called ***optimal control problems,*** but these are difficult to solve. Consequently, they are often simplified by reformulating them into static optimization problems.

Classification according to the nature of a solution

Global optimization problems—we are looking for a global optimum (global extremum) of the optimization problem (2.1). Therefore, these problems are more complex and generally computationally difficult.

Local optimization problems—we are looking for a local optimum (local extremum) of an objective function of the optimization problem (2.4). In special cases, when the objective function has a single extremum in a feasible region, it is also a global extremum (for a convex or concave objective function). We only need to focus on a simpler *problem of local optimization,* i.e., to search for a local extremum.

Classification according to the number of variables of an objective function

One-dimensional optimization—the objective function is a function of one real variable. One-dimensional optimization plays an important role in solving optimization problems for functions of several variables. The issue of determining the minimum value of the so-called *unimodal function,* i.e., a function that has a single minimum at a specified closed interval $[a, b]$.

Multidimensional optimization—the objective function is a function of several variables. Many of multidimensional optimization methods use one-dimensional minimization to determine a minimum value of a function in each direction.

Classification by number of objective functions (number of optimization criteria)

Single criterion optimization—an optimization problem is expressed by one objective function and appropriate constraints, i.e., a system of equations and inequalities. The aim of single criterion optimization is to find the optimum (in the sense of extremalization) of one objective function under given conditions (constraints). For the case of single criterion minimization, the problem is not solvable if a feasible region is empty or if the objective function is not lower bound. Analogously, for the problem of single criterion maximization, we consider the upper bound.

Multicriteria optimization—solving an optimization problem often leads us to explore multicriteria objective functions. The relationships between the considered criteria may be different but not necessarily of the same size and their importance may not be the same. Therefore, the criteria must be necessarily preferred or indifferent to each other and, in addition, the criteria must be comparable [4].

Of course, there are other ways and criteria for allocating optimization problems. It is advisable to have an overview of individual optimization problems as their solving methods are also based on the type of optimization problems. Sometimes problems can be solved analytically, i.e., by deriving relationships from which the exact solution can be calculated. Then we use the so-called *indirect methods* [3, 5, 14, 20, 25, 33, 34].

However, when solving practical optimization problems, very often such a solution is not possible due to the complexity of a problem. Therefore, *numerical (direct) methods of solution* are used to solve practical optimization problems. As we know, application of optimization in practice has a wide scope. Later, we will focus mainly on methods applicable to the optimization of technological processes.

In the following sections, we will briefly get acquainted with the most frequently used types of optimization problems considering their application in engineering optimization. We will not present their theoretical basis as this is not the aim of this monograph. Later, we will discuss in more detail the selected *optimization methods* that are suitable or predestined for solving engineering optimization problems. Sometimes, we will refer to the appropriate literature.

2.3.1 Linear Programming (LP)

Liner programming is one of the most common problems in mathematical programming. The objective of linear programming (LP) is called the problem of *minimizing* (*maximizing*) a linear function at given linear constraints [11, 14, 23, 32, 34, 35].

The LP problem is usually written in one of the following standard forms

$$Min\left\{\boldsymbol{c}^T\boldsymbol{x}\ |\boldsymbol{A}\boldsymbol{x} \geq \boldsymbol{b}, \boldsymbol{x} \geq \boldsymbol{0}\right\}, \tag{P1}$$

$$Min\left\{\boldsymbol{c}^T\boldsymbol{x}\ |\boldsymbol{A}\boldsymbol{x} \geq \boldsymbol{b}\right\}, \tag{P2}$$

$$Min\left\{\boldsymbol{c}^T\boldsymbol{x}\ |\boldsymbol{A}\boldsymbol{x} = \boldsymbol{b}, \boldsymbol{x} \geq \boldsymbol{0}\right\}. \tag{P3}$$

where $\boldsymbol{A} \in \mathbb{R}^{m \times n}$, $\boldsymbol{b} \in \mathbb{R}^m$, $\boldsymbol{c} \in \mathbb{R}^n$

We will call the problems (P1)–(P3) the ***problems of linear programming in a primary form***. The sets

$$P_1 = \{\boldsymbol{x}\ |\boldsymbol{A}\boldsymbol{x} \geq \boldsymbol{b}, \boldsymbol{x} \geq \boldsymbol{0}\},$$

$$P_2 = \{\boldsymbol{x}\ |\boldsymbol{A}\boldsymbol{x} \geq \boldsymbol{b}\},$$

and

$$P_3 = \{\boldsymbol{x}\ |\boldsymbol{A}\boldsymbol{x} = \boldsymbol{b}, \boldsymbol{x} \geq \boldsymbol{0}\}$$

will be called *a feasible region of the (primary) problem* (P1), (P2), and (P3), respectively. The feasible region of the LP problem is the intersection of a finite number of closed affine half-spaces, thus it is polyhedral. Each LP problem can be converted into any of the forms (P1)–(P3). Therefore, in the literature, some statements are formulated only for the problem (P1), although by analogy they remain valid for the problems in the form (P2), (P3).

Theorem 2.3.1 (Basic theorem of linear programming)

For each linear programming problem (P1) there is exactly one of the following possibilities:

1. $\exists \boldsymbol{x}^* \in \mathcal{P}_1 : \quad \boldsymbol{c}^T \boldsymbol{x}^* \leq \boldsymbol{c}^T \boldsymbol{x} \quad \forall \boldsymbol{x} \in \mathcal{P}_1$ *–optimality,*
2. $\inf_{\boldsymbol{x} \in \mathcal{P}_1} \boldsymbol{c}^T \boldsymbol{x} = -\infty$ *–infinity,*
3. $\mathcal{P}_1 = \emptyset$ *–infeasibility.*

The problem of LP in the form (P3) is used in the interior point methods presented by the Indian mathematician Narendra Karmakar in 1984. He has caused a phenomenon called the revolution in optimization [36, 37]. In general, we can convert each LP problem with the constraints given by a system of linear equations and inequalities to an equivalent problem with linear equations using additional variables. We call the form (P3) *the standard form of linear programming problem.* We can find several versions of its notation in the literature, more extensive *scalar notation of LP*, *vector notation of LP* or frequently used *matrix form of LP*:

$$\begin{aligned} f_0(\boldsymbol{x}) = \boldsymbol{c}^T \boldsymbol{x} &\to \min \\ \boldsymbol{A}\boldsymbol{x} &= \boldsymbol{b} \qquad \text{(SLP)} \\ \boldsymbol{x} &\geq 0 \end{aligned}$$

where $f_0(\boldsymbol{x}) = c_1 x_1 + c_2 x_2 + \cdots + c_n x_n$ is a linear objective function of several variables, $\boldsymbol{c} = (c_1, c_2, \ldots c_n)^T$ is a vector of coefficients of the objective function, elements a_{ij} of the matrix $\boldsymbol{A} \in \mathbb{R}^{m x n}$ are called constraint coefficients, vector coordinates $\boldsymbol{b} = (b_1, b_2, \ldots, b_m)^T$ are called the coefficients of the right-hand side of a constraint. A matrix $\boldsymbol{A} \in \mathbb{R}^{m x n}$ and a vector $\boldsymbol{b} \in \mathbb{R}^m$ are given, $\boldsymbol{x} \in \mathbb{R}^n$ is a vector variable.

It will be appreciated that each minimizing problem of LP in the form (P1), (P2) or (P3) is associated with a uniquely identified maximizing problem having certain advantageous and useful properties. We call such a problem a *dual problem of linear programming* (D1), (D2), (D3) belonging to the primary problem (P1), (P2), (P3).

For example, the dual problem (D1): $Max\{\boldsymbol{b}^T \boldsymbol{y} \mid \boldsymbol{A}^T \boldsymbol{y} \leq \boldsymbol{c}, \boldsymbol{y} \geq \boldsymbol{0}\}$, with a feasible region (of the dual) problem $D_1 = \{\boldsymbol{y} \mid \boldsymbol{A}^T \boldsymbol{y} \leq \boldsymbol{c}, \boldsymbol{y} \geq \boldsymbol{0}\}$ belongs to the primary problem (P1).

When developing appropriate algorithms in linear programming, it has proved appropriate to use both duality and complementarity, i.e. the so-called *weak duality theorem* and its consequences and a *strong duality theorem* or *complementarity theorem.* See more in [14, 22].

So far, the best known and successfully used algorithms for solving LP problems are the *simplex method* and the *interior point method* discussed in Chap. 4 that is devoted to the selected methods of multidimensional optimization. Modern algorithms give us the opportunity to solve linear programming problems with many variables. Linear programming also plays an important role in nonlinear optimization problems. Some mathematical problems formulated as nonlinear can be linearized after additional modification and then, thanks to good implementation and availability of suitable algorithms, solved using linear programming. In Chap. 6 we will discuss the problem of optimizing cutting conditions during machining using the linear formulation of the problem (after the transformation).

2.3.2 *Quadratic Programming (QP)*

Quadratic programming is one of the other types of frequently used optimization problems [35]. The objective function f_0 is quadratic, but constraint $f_i,\ i \in I,\ h_j,\ j \in J$ are linear. Literature introduces a notation for the standard form of the quadratic programming problem as follows

$$\begin{aligned} f_0(\boldsymbol{x}) = \tfrac{1}{2}\boldsymbol{x}^T\boldsymbol{Q}\boldsymbol{x} + \boldsymbol{c}^T\boldsymbol{x} &\rightarrow \min \\ \boldsymbol{A}\boldsymbol{x} &= \boldsymbol{b} \\ \boldsymbol{x} &\geq 0 \end{aligned} \qquad \text{(SQP)}$$

where $\boldsymbol{A} \in \mathbb{R}^{mxn}$, $\boldsymbol{b} \in \mathbb{R}^m$, $\boldsymbol{c} \in \mathbb{R}^n$, $\boldsymbol{Q} \in \mathbb{R}^{nxn}$ are known, and $\boldsymbol{x} \in \mathbb{R}^n$ is a vector variable. The expression $\frac{1}{2}\boldsymbol{x}^T\boldsymbol{Q}\boldsymbol{x}$ represents a quadratic form of the objective function of several variables. In this model, the constraints are not limited only to equality. A notation can also contain linear inequalities, which, however, we convert to equations using additional variables and thus get this standard form. Without prejudice to the generality, we can assume that a matrix $\mathbf{Q}$ is symmetrical because $\boldsymbol{x}^T\boldsymbol{Q}\boldsymbol{x} = \frac{1}{2}\boldsymbol{x}^T(\boldsymbol{Q} + \boldsymbol{Q}^T)\boldsymbol{x}$. If symmetrical matrix $\mathbf{Q}$ is positive definite, then then the objective function is convex. A suitable numerical method for solving a quadratic programming problem with a positive semidefinite matrix $\boldsymbol{Q}$ is *the interior point method* for quadratic programming.

2.3.3 *Nonlinear Programming (NP)*

If in the mathematical programming problem (2.6) that is now called (NP)

$$Min\{f_0(\boldsymbol{x})\ |\boldsymbol{x} \in X, f_i(\boldsymbol{x}) \leq 0, i \in I,\ h_j(\boldsymbol{x}) = 0,\ j \in J\} \quad \text{(NP)}$$

at least one of the functions $f_0, f_i,\ i \in I,\ h_j,\ j \in J$ is not linear and a set of feasible solutions has the following form:

$$K = \{x \in \mathbb{R}^n \mid f_i(x) \le 0, i \in I,\ h_j(x) = 0,\ j \in J\},$$

then we talk about *nonlinear programming with constraints.*

Let us suppose that $\emptyset \ne X \subseteq \mathbb{R}^n$. The sets I, J are index sets, i.e. $I = \{1, 2, \ldots, m\}$ and $J = \{m+1, m+2, \ldots, m+r\}$, there is no relationship between the number of constraints and the number of variables n. Let us present the classification of nonlinear programming problems from a historical point of view.

If K is an open set, then it is an ***unconstrained nonlinear optimization problem*** (also known as free boundary extremum problem)

$$Min\{f_0(x) \mid x \in X = \mathbb{R}^n\} \quad \text{(U1)}$$

If $K = \{x \in \mathbb{R}^n \mid h_j(x) = 0,\ j \in J\}$ *is a closed set described by a system of equations, then it is* ***an constrained nonlinear optimization problem***.

$$Min\{f_0(x) \mid x \in X,\ h_j(x) = 0,\ j \in J\} \quad \text{(U2)}$$

If $K = \{x \in \mathbb{R}^n \mid f_i(x) \le 0, i \in I\}$ *is a closed set described by a system of inequalities, then it is a* ***problem of nonlinear programming in the narrower sense***

$$Min\{f_0(x) \mid x \in X,\ f_i(x) \le 0, i \in I\} \quad \text{(U3)}$$

If $K = \{x \in \mathbb{R}^n \mid f_i(x) \le 0, i \in I,\ h_j(x) = 0,\ j \in J\}$ *is a closed set described by a system of equations and inequalities, then it is a* ***problem of nonlinear programming in the broader sense***

$$Min\{f_0(x) \mid x \in X, f_i(x) \le 0, i \in I,\ h_j(x) = 0,\ j \in J\} \quad \text{(U4)}$$

Let us denote K_1, K_2, K_3, K_4 the sets of feasible solutions for the corresponding problems.

(U1)—is the simplest case of a NP problem because we have no limits (boundaries), it is an unconstrained nonlinear optimization problem. This problem is associated with the very origin of the differential calculus, so it naturally belongs to serial number one.

(U2)—it is the classical (Lagrange) problem to determine extreme values of a function subject to constraints. In 1762, French mathematician Joseph Louis Lagrange [23, 33] was the first to formulate and solve a constrained nonlinear optimization problem in which the constraints were in the form of equations. He devised *the method of Lagrange Multipliers* for finding minimum or maximum values of a function subject to constraints. Historically, this is the second problem of the NP.

(U3)—In 1951, American mathematicians H. W. Kuhn and A. W. Tucker analysed the problem of nonlinear programming (U3) and published their optimality conditions known as the Kuhn-Tucker conditions. Later it turned out that the necessary conditions of extremes with boundaries such as inequalities were published by the American mathematician W. Karush earlier in his diploma thesis, so we know them today under the name Karush–Kuhn–Tucker conditions. See more in [14, 23, 24, 26, 38].

(U4)—is a nonlinear programming problem that is a combination of the problems (U2) and (U3). The problems (U1)–(U4) are usually analysed using a differential calculus.

2.4 Optimality Conditions

Optimality conditions are helpful when examining the solvability of unconditional optimization problems (unconstrained optimization problems) in terms of finding a local extremum. We will introduce optimality conditions for minimization problems. The basic division of optimality conditions is according to the order of derivatives of a given function. Further categorization can occur if we decide whether this is a necessary or sufficient condition [19].

Theorem 2.4.1 (First order necessary conditions for optimality)
Let $f \in C^1$ and let $\boldsymbol{x}^*$ be a point of a local minimum of a function f. Then at $\boldsymbol{x}^*$ the following is valid

$$\nabla f\left(\boldsymbol{x}^*\right) = g\left(\boldsymbol{x}^*\right) = \boldsymbol{0}, \tag{2.10}$$

i.e. $\boldsymbol{x}^*$ is a stationary point of a function f.

Theorem 2.4.2 (Second order necessary conditions for optimality)
Let $f \in C^2$ and let $\boldsymbol{x}^*$ be a point of a local minimum of a function f. Then at $\boldsymbol{x}^*$ the following is valid

1. $g(\boldsymbol{x}^*) = \boldsymbol{0}$,
2. $\boldsymbol{H}(\boldsymbol{x}^*)$ *is positive semidefinite matrix, i.e.*

$$\boldsymbol{p}^T \boldsymbol{H}\left(\boldsymbol{x}^*\right) \boldsymbol{p} \geq 0 \qquad \forall \boldsymbol{p} \in \mathbb{R}^n \tag{2.11}$$

Theorem 2.4.3 (Second order sufficient conditions for optimality)
Let $f \in C^2$ and let at point $\boldsymbol{x}^*$ it holds

1. $g(\boldsymbol{x}^*) = \boldsymbol{0}$,
2. $\boldsymbol{H}(\boldsymbol{x}^*)$ *is positive semidefinite matrix, i.e.*

$$\boldsymbol{p}^T \boldsymbol{H}\left(\boldsymbol{x}^*\right) \boldsymbol{p} > 0 \qquad \forall \boldsymbol{p} \in \mathbb{R}^n, \boldsymbol{p} \neq 0 \tag{2.12}$$

Then a point $\boldsymbol{x}^*$ is a point of sharp local minimum of a function f.

We have used the terms defined in Sect. 2.1, thus $\boldsymbol{p}$ is a direction vector,$\boldsymbol{H}(\boldsymbol{x}^*)$ is the Hessian of a function f at a point $\boldsymbol{x}^* \in \mathbb{R}^n$.

2.5 Engineering Optimization (Process Optimization)

Optimization in the broadest sense is a process of getting the best results under certain conditions. In the field of design, construction or operation of any engineering system, engineers make many technological and managerial decisions on several levels and stages. Decisions are made to meet the required criteria. The goal of all such decisions is to minimize the effort or cost expended, or to maximize the required benefit. Since we can express the effort or required benefit as a criterion function of certain decision variables, engineering optimization can be defined as a process of finding conditions that ensure maximum or minimum values of a criterion function.

The purpose of engineering optimization is to provide tools for deciding on the best solution for specific problems which can be of various natures (e.g. to determine optimal cutting conditions to achieve the prescribed quality of the machined surface; to set stable values of production parameters to minimize production costs while maintaining technological procedures; to establish a plan for products distribution from a network of warehouses in order to minimize time and cost of delivery, etc. Several steps are needed to make such decisions competently.

Sequence of steps when solving an optimization problem.

According to literature resources [3, 30, 31, 39–42], in optimal decision making (optimal control) it is necessary to know:

(a) mathematical model of a control object
(b) objective function
(c) constrained conditions.

Mathematical model. Its creation is the first step on the way to an optimal solution. It is necessary to create a corresponding model that mathematically describes the relevant object, system, or process under investigation. For the meaningful decision-making process, its object must have defined *quantifiable parameters*:

- *input parameters* that we can influence,
- *output parameters* that we want to influence.

The model itself, as a mathematical description of a system, must contain a *quantitative criterion* (profit, cost, surface roughness, etc.), based on which it will be possible to assess the success of parameter settings. It must be possible to determine a value of the criterion by measurement or calculation.

Constraint conditions. To be as close as possible to reality, a mathematical model can contain various constraints. During an optimization process, we avoid infeasible solutions to a given problem. The success of the whole optimization problem depends on the creation of an adequate model. If a model is too simplified, the results may not correspond to reality. If we create a model that is too complex, solving a problem becomes complicated or even impossible. Optimization problems are often solved during a creation of a model itself in identifying input–output dependencies.

Having created a model, it is possible to *formulate an optimization problem.* Let us select a suitable type of a problem.

Objective function. Its formulation is a key step to proper optimization and its selection requires *deep engineering experience* in the issue of optimization problems. It must be based on the very essence of a model, which as a mathematical description, and must contain a *quantitative criterion.* In addition, the form of an objective function applicable for any optimization problem from engineering area is suitable for optimization only under certain conditions and it is not a generally valid form.

The selection of a suitable optimization algorithm and the solution of an optimization problem follows the implementation of the previous steps. As mentioned above, by choosing an optimization method, a user influences both computational and time-consuming aspects as well as the success of the entire optimization process. It is necessary to provide a suitable software in advance. Before optimization we must verify the correct operation of the programmed methods. It is recommended to use at least two testing n–dimensional functions for testing: the Rosenbrock function and its modifications, which will be described in the next chapters.

Many *computer programs* are available to solve nonlinear programming problems. A summary overview of the availability of some optimization packages within software systems can be found in [11, 43]. We will mention the software system MATLAB, Mathematica or FOSS functions of the program QtOctave [44]. During a calculation process, it is important to realize that a method that works well for one type of optimization problems may work poorly for others. Therefore, it is usually necessary to try more than one method to effectively solve a particular problem. In addition, the effectiveness of any nonlinear programming method depends to a large extent on values of adjustable parameters, such as an initial point, a step length and convergence requirements. The correct set of values for these adjustable parameters can be obtained either by trial and error or through experience gained when working with a method to solve similar problems. It is therefore desirable to run a program with different initial points to avoid a local or false optimum in global optimization [2, 11, 14, 25, 33, 45].

Verification of the results. Having obtained a solution using one of the optimization techniques, the result must be evaluated to verify whether this is the appropriate solution to the optimization problem [5]. Optimality conditions are used to verify a solution. If the obtained solution does not meet the conditions of optimality, it is necessary to continue the search. Useful information can be provided by an analysis of a solution based on evaluation of optimality conditions. Finally, the interpretation of a solution from the point of view of a specific problem is important. This may lead to a possible modification (refinement) of a model. Sensitivity analysis of a model is also useful (an assessment of how a solution changes when changing model parameters).

Optimization has its application in processes control at all levels: at the level of elementary processes, technological processes, production processes as well as in planning processes of strategic importance. We encounter the issue of optimization in complex production planning at the level of multinational companies, individual operations, transport optimization, technological processes design, stock optimization, etc.

The optimization problem (2.2) is an abstraction of a problem of making the best possible vector selection in $\mathbb{R}^n$ from a set of suitable candidates (feasible variants of a solution). A variable $\boldsymbol{x}$ represents the realized choice of a solution, constraints $f_i(\boldsymbol{x}) \leq \boldsymbol{b}_i$ represent fixed requirements or specifications that must be respected when making decisions (they limit a possible choice) and an objective function $f_0(\boldsymbol{x})$ represents the costs when selecting $\boldsymbol{x}$. (When changing a sign $-f_0(\boldsymbol{x})$ it represents revenue, profit, or benefit of selection $\boldsymbol{x}$). The solution of the optimization problem (2.2) then corresponds to the selection of minimum costs (or maximum profit) from all the selection of variants that meet the defined requirements. In other words, optimization finds its application, for example in the economy to maximize profits or minimize costs and risk. Buyers try to reach an optimal decision when making decisions on price, quality, and availability.

Examples of optimization can be found in almost all areas of human activity and decision making. For example, in logistics we want to minimize transport costs and time, in industrial production it is desirable to minimize costs and production time under given conditions while maintaining the required quality. Engineers optimize parameters of proposed objects: control system or controlled object, optimize technological procedures to determine optimal conditions during machining as well as cutting processes. Even in medicine and in nature, we can find application for optimization (natural laws can be interpreted as an "effort" of a system to reach a state with minimal energy).

Many of these problems can be formulated in a uniform way as an optimization problem and appropriate mathematical methods and algorithms can be used to solve such a problem. From a *mathematical point of view*, it is a process of finding the extrema of an objective function.

References

1. Avriel, M., Rijckaert, M. J., & Wilde, D. J. (1973). *Optimization and design* (2nd ed., p. 489). Cambridge University Press, Prentice Hall.
2. Messac, A. (2015). *Optimization in practice with MATLAB for engineering students and professionals* (496 p.). Cambridge University Press. ISBN 978–1–107–10918–6.
3. Kostúr, K. (1991). *Optimalizácia procesov* (p. 365). Technická univerzita v Košiciach.
4. Kažimír, I., & Beňo, J. (1989). *Teória obrábania. Návody na cvičenia* (280 p.). Alfa.
5. Rosinová, D., & Dúbravská, M. (2008). *Optimalizácia.* STU v Bratislave. 195 s. ISBN 978-80-227-2795-2.
6. Afanasiev, V. N., Kolmanovskii, V. B., & Nosov V. R. (1996). *Mathematical theory of control systems design.* ISBN 978–94–017–2203–2.
7. Antoniou, A., & Lu, W. S. (2007). *Practical optimization. Algorithms and engineering applications* (675 p.). Springer Science & Business Media LCC. ISBN-13: 978-0-387-71106-5.
8. Bazaraa, M. S., Sherall, H. D., & Shetty, C. M. (1993) *Nonlinear programming. Theory and algorithms* (2nd ed.). Wiley.
9. Bertsimas, D., & Tsitsiklis, J. N. (1997). *Introduction to linear optimization* (p. 608). Athena Scientific.
10. Loladze, A. T. (1989). *Základy optimalizácie strojárskej technológie* (216 p.). Alfa. ISBN 80–05–00083–9.

11. Rao, S. (2009). *Engineering optimization. Theory and practice* (4th ed., 830 p.). John Wiley & Sons, Inc.
12. Реклейтис, Г., Рейвиндран, А., & Рэгсдел, К. (1986). *Оптимизация в технике* 1, [Reklaitis, G.V., Ravindran, A. Ragsdell, K. M. *Engineering optimization . Methods and applications*] (350 p.). Mir, Moskva.
13. Реклейтис, Г., Рейвиндран, А., & Рэгсдел, К. (1986). *Оптимизация в технике* 2, [Reklaitis, G.V., Ravindran, A., & Ragsdell, K. M. *Engineering optimization. Methods and applications*] (320 p.). Mir, Moskva.
14. Boyd, S., & Vandenberghe, L. (2009). *Convex optimization* (701 p.). Cambridge University Press. ISBN 978-0-521-83378-3.
15. Hliník, J. (2015) *Tvarová optimalizace difuzoru vodní turbíny*. VUT v Brně, Brno. 55 p. (Bachelor's thesis).
16. Ivan, J. (1989). *Matematika 1* (704 p.). Alfa.
17. Ivan, J. (1989). *Matematika 2* (632 p.). Alfa. ISBN 80–05–00114–2.
18. Jarník, V. (1976). *Diferenciální počet I*. Academia.
19. Machalová, J., & Netuka, H. (2013). *Numerické metody nepodmíněné optimalizace* (142 p.). 1. vyd. Univerzita Palackého v Olomouci, Olomouc, Czech Republic. ISBN 978–80–244–3403–2.
20. Marčuk, G. I. (1987). *Metódy numerické matematiky* (528 p.). Academia, nakladatelství ČSAV.
21. Rektorys, K., et al. (1981) *Přehled užité matematiky*. SNTL. 1140 s.
22. Brunovská, A. (1990). *Malá optimalizácia* (248 p.). Alfa. ISBN 80-05-00770-1.
23. Hamala, M., & Trnovská, M. (2012). *Nelineárne programovanie/Nonlinear programming* (339 p.). Epos. ISBN 978-80-805-7986-9.
24. Lenstra, J. K., Rinnooy, K., & Schrijver, A. (1991). *History of mathematical programming. A collection of personal reminiscences* (141 p.). Elsevier Science Pubications CWI.
25. Yang, W.Y., Cao, W., Chung, T.S., & Morris, J. (2005). *Applied Numerical Methods Using MATLAB* (p. 509.). John Wiley & Sons, Inc.
26. Hamala, M. (1972). *Nelineárne programovanie*. Alfa.
27. Cao, H., Qian, X., & Zhou, Y. (2018). Large-scale structural optimization using metaheuristic algorithms with elitism and a filter strategy. *Structural and Multidisciplinary Optimization, 57*, 799–814 (2018). https://doi.org/10.1007/s00158-017-1784-3.
28. Čermák, L., & Hlavička, R. (2016). *Numerické metódy. Numerical methods* (110 p.). CERM, Brno. ISBN 978-80-2145437-8.
29. Dupačová, J., & Lachout, P. (2011). *Úvod do optimalizace* (81 p.). Matfyzpress. ISBN 978-80-7378-176-7.
30. Hudzovič, P. (2001). *Optimalizácia* (320 p.). STU v Bratislave. ISBN 80–227–1598–0.
31. Panda, A., Duplák, J., Jurko, J., & Pandová, I. (2013). Roller bearings and analytical expression of selected cutting tools durability in machining process of steel 80MoCrV4016. *Applied Mechanics and Materials. Automatic Control and Mechatronic Engineering, 415*, 610–613. ISSN 1660-9336.
32. Plesník, J., Dupačová, J., & Vlach, M. (1990). *Lineárne programovanie*. Alfa.
33. Stewart, J., & Clegg, D. (2012) *Brief applied calculus* (491 p.). BROOKS/COLE Cengage Learning, Canada. ISBN 978-1-111-57005-7.
34. Vanderbei, R. J. (2008). *Linear programming: Foundations and extensions* (3rd ed.). Springer.
35. Ban, N., & Yamazaki, W. (2021). Efficient global optimization method via clustering/ classification methods and exploration strategy. *Optimization and Engineering, 22*, 521–553 (2021). https://doi.org/10.1007/s11081-020-09529-4.
36. Nesterov, Y. E., & Nemirovski, A. S. (1994). *Interior point polynomial algorithms in convex programming*. SIAM Publications.
37. Halická, M. (2004). Dvadsať rokov moderných metód vnútorného bodu. *Pokroky matematiky, fyziky a astronomie, 49*(3), 234–244.
38. Колманоьский, В. Б. (1997). *Задачи оптимального управления. Математика*. [Kolmanovskii, V. B. Problems of Optimal Control. Mathematics] (pp. 121–127). Московский государственный институт електроники и математики, Москва.

39. Hudzovič, P. (1990). *Identifikácia a modelovanie/Identification and modeling* (2nd ed., 255 p.). Slovenská vysoká škola technická v Bratislave. ISBN 80–227–0213–7.
40. Jurko, J., Džupon, M., Panda, A., & Zajac, J. (2012). Study influence of plastic deformation a new extra low carbon stainless steels XCr17Ni7MoTiN under the surface finish when drilling. *Advanced Materials Research: AEMT 2012, 538–541*, 1312–1315.
41. Jurko, J., Panda, A., Gajdoš, M., & Zaborowski, T. (2011). Verification of cutting zone machinability during the turning of a new austenitic stainless steel. In *Advances in Computer Science and Education Applications: International Conference CSE 2011* (pp. 338–345). Springer.
42. Plevný, M., & Žižka, M. (2010). *Modelování a optimalizace v manažerském rozhodování*. Západočeská univerzita v Plzni, Plzeň. 2. vydání. 298 s. ISBN 978-80-7043-933-3.
43. Moré, J. J., & Wright, S. J. (1993). *Optimization software guide*. Society of Industrial and Applied Mathematics.
44. http://qoctave.wordpress.com/.
45. Schrijver, A. (2017). *A course in combinatorial optimization* (p. 221). University of Amsterdam.

Chapter 3
Optimization Methods in General

As we have mentioned above, there is no one universal method for different types of optimization problems. We have a set of optimization methods (algorithms) suitable for individual types of problems. Many factors affect how an optimization problem can be effectively solved.

The number of equations and inequalities that define a feasible region X, $|I|+|J|$ is one of them. Properties of objective and constraint functions f_0, f_i, h_j have an important influence on complexity of calculation, especially continuity, (non)linearity, differentiability, convexity of f_0, convexity of the set X, etc. For this reason, several methods, and different types of algorithms to solve different optimization problems have been developed.

It is up to a user to decide which method to use for a given optimization problem. By choosing a suitable optimization method, we can affect the efficiency of the entire optimization process, its success and computational complexity. With end-user engineering optimization, we are interested in what facts need to be considered when choosing a specific method to solve a given optimization problem. Some of them are as follows [1]:

- type of optimization problem that we are going to solve (the problem of nonlinear programming, geometric programming, etc.)
- availability of software for a given optimization method (already developed program)
- time needed to develop a program to solve a problem
- whether a method requires derivatives of an objective function f
- knowledge of the effectiveness of a method
- accuracy of a required solution
- robustness and reliability of a method in finding an optimal solution
- generality of a program (applicability) to other optimization problems
- ease (or complexity) of using a program as well as possibilities of outputs interpretation
- engineering experience in solving a problem.

A. Vagaská et al., *Optimization Methods in Mathematical Modeling of Technological Processes*, Mathematical Engineering, https://doi.org/10.1007/978-3-031-35339-0_3

The overview and classification help us to orient ourselves in the system of optimization methods.

3.1 Classification of Optimization Methods

Optimum search methods are also known as *mathematical programming methods* and are generally studied as part of *operations research (operational research)* which is one of the fields of applied mathematics focused on the use of mathematical models, statistics and algorithms in decision making to achieve an optimal solution. Table 3.1 presents various optimization methods together with other known areas of operations research. This list is not unique, it is given for the sake of clarity [1].

Mathematical programming methods are useful for finding a minimum of a function of several variables with prescribed constraints. Stochastic process methods are used to analyse problems described by a set of random variables with a known probability distribution curve. Statistical methods make it possible to analyse experimental data and create empirical models to obtain the most adequate model of a system.

Different methods for solving various types of optimization problems are presented in Table 3.1 in the first column. Classical methods of differential calculus

Table 3.1 *Operations research methods* [1]

Mathematical programming or optimization methods	Methods of stochastic processes	Statistical methods
Differential calculus methods *Variation number* *Nonlinear programming* *Geometric programming* *Quadratic programming* *Linear programming* *Dynamic programming* *Integer programming* *Stochastic programming* *Discrete programming* *Multicriteria optimization* *Network methods:* *CPM and PERT* *Game theory*	*Statistical decision theory* *Markov processes* *Renewal theory* *Queueing theory* *Simulation methods* *Reliability theory*	*Regression analysis* *Cluster analysis* *Pattern recognition* *Design of experiments* *Discriminant analysis (factor analysis)*
Modern and non-traditional optimization techniques *Genetic algorithm* *Simulated annealing* *Ant colony optimization* *Particle swarm optimization* *Neural networks* *Fuzzy optimization*		

are used to find the extrema of a function of several variables (*unconstrained optimization problem*). These methods work on the assumption that a function is differentiable and has continuous second order partial derivatives. For optimization problems with equality constraints, the *method* of *Lagrange multipliers* is widely *used* to solve *constrained optimization problems*. For problems with inequality constrains, the Karusha-Kuhn-Tucker conditions are used to identify an optimal solution. However, these methods lead to a system of nonlinear simultaneous equations, which can be computationally demanding. From the names of the methods of nonlinear, linear, geometric or integer programming we can indicate the type of optimization problems and the methods to be used. Many of them are numerical methods, which are characterized by a fact that approximation process of an optimal solution is realized in an iterative way from some initial point. The dynamic programming method is also a numerical procedure useful primarily for solving problems of optimal control. Stochastic programming deals with solving optimization problems in which some variables are described by probability distributions [1].

Let us introduce classification of optimization methods and some classification aspects.

Exact solution & approximate solution

From the point of view of what optimal solution we achieve using a given method, we can divide optimization methods into two basic types: analytical and numerical.

Analytical optimization methods—are also called *indirect methods* and provide us to find the *exact value* of an optimal solution $\boldsymbol{x}^*$ using exact or analytical procedure by deriving relationships from which the exact solution of a problem can be calculated. However, this is only possible in some cases. Very often, in practical optimization problems due to their complexity such a solution is not possible. Therefore, approximate methods are used more often.

Numerical optimization methods are sometimes called *direct methods*. They allow us to find an approximate solution to an optimization problem with the required accuracy. Most practical optimization problems need to be solved *numerically* using *iterative methods* where we get only an approximation of a searched point $\boldsymbol{x}^*$.

In engineering optimization, we focus mainly on numerical methods. Therefore, the next chapter will deal with *approximate solution methods* and computational algorithms.

Computational algorithms, which are designed to minimization, are divided into

- ***deterministic*** and ***stochastic*** depending on whether they use random decision making
- ***interior point*** algorithms ($\boldsymbol{x}^* \in K$) and ***exterior point*** algorithms ($\boldsymbol{x}^* \notin K$) depending on whether they construct or do nor construct a sequence of approximate solutions that are permissible solutions to a given problem
- ***direct*** (functional values), ***gradient*** (first derivatives for gradients), and ***Newton's*** (matrices of second derivatives) depending on the higher order derivative an algorithm uses.

Selection and compilation of an algorithm for a particular optimization problem must consider the type of optimization problem, heuristics, the form of an algorithm and convergence properties of an algorithm. When minimizing an objective function, we need to find the point of a local minimum, therefore the advantage is convexity and polyhedral feasible region and it is essential to perceive the structure of an optimization problem [2, 3].

Boundedness & unboundedness

Another important aspect in the classification of optimization methods. They are as follows

Numerical methods for solving unconstrained minimization problems

- using first and second derivatives (the Newton–Raphson method)
- using first derivatives (e.g. gradient methods, methods of combined directions, quasi-Newton methods)
- direct (comparative) methods of minimization without constraints
  - with deterministic search (e.g. the Hooke–Jeeves method, the Nelder-Mead method, the Box-Wilson method, the Gauss-Seidl method)
  - with random search (e.g. genetic algorithms, simulated annealing)

Numerical methods for constrained minimization problems

- inequality constraints (e.g. method of possible directions, the barrier functions method, penalty method, etc.)
- equality constraints (e.g. the Lagrange multiplier method, reduced gradient method, the Newton's method of reduced Hessian matrix, penalty method, etc.)

Depending on the order of derivation of an objective function f the optimization method uses, we consider the following ***types of methods***

- zero (comparative, relaxing)
- one (gradient methods)
- two (Newton's gradient methods)

or they are referred to as zero, first or second order methods.

Number of variables of an objective function f

Another important criterion for optimization methods classification is the fact whether it is minimization of a function of one variable or several variables. Then, we classify them as follows:

One-dimensional optimization methods (*methods for minimizing a function of one variable*).

Multidimensional optimization methods (*methods for minimizing a function of* n-variables).

This classification reflects the fact that minimizing a function of one variable is a qualitatively simpler task than that of a function of n variables. The task of minimizing a function of one variable is the cornerstone for many iterative methods of solving multidimensional optimization problems in which the need to solve one-dimensional minimization problem occurs in each iteration. We will explain the reason and in the following sections we will focus on optimization methods regarding this classification criterion.

With constrained optimization, we apply the fact that constrained ***optimization problem*** of the function of n-variables can be transformed to solve sequences of partial unconstrained problems for searching the extreme value, which is much easier to solve when compared to the original problem. However, if we want to be successful, the assumption to solve an unconstrained optimization problem effectively must be met. The problem of unconstrained optimization of a function of we apply the fact that constrained ***optimization problem*** of the function of n-variables is still relatively complex problem that needs to be solved using some of the ***iterative methods*** while individual iterations require further solutions of simpler subproblems.

The largest class of numerical methods for solving free optimisation problems consists of methods, one iteration of which can be expressed as follows:

$$point\, x^k \rightarrow direction\, s^k \rightarrow step\, \lambda_k$$

In more detail, in the given iteration we have some approximate solution, i.e., we have an initial *point* x^k of n-dimensional space. At this *point* x^k we set a *direction* s^k, i.e. nonzero n-dimensional vector ($s^k \in \mathbb{R}^n$), in which a minimized function $F(x)$ will show a decrease. Therefore, the term *descent direction* is used. Finally, in each *direction* s^k we realize the *step* λ_k, thus a shift from original to a new point, *step* λ_k we express with a positive number. This completes one iteration and the whole procedure can be repeated. In other words, within such an iterative method, minimum value of a relevant function $F(x)$ is searched around an initial point by gradually "shifting" in each iteration (iteration step) to the point with the "better" value of a function $F(x)$, until required tolerance for a solution is reached. We understand tolerance as a value determining a criterion of completing calculation (e.g., change of a functional value compared to a previous step, or a size of a gradient in gradient methods, etc.).

The formal notation of the iteration scheme has the following form:

$$x^{k+1} = x^k + \lambda_k s^k, \quad (k = 0, 1, 2, \ldots). \tag{3.1}$$

Individual *numerical methods* differ from each other by rules of selection *direction* s^k and *step* λ_k. In some numerical methods, direction s^k is usually defined by some simple formulas. For example, with the steepest descent method, direction s^k is defined as a negative gradient of a minimized function $F(x)$ at a point x^k. In the Definition 2.1.6a we have denoted the descent direction as a vector $\boldsymbol{p} \in \mathbb{R}^n$, in iterative methods in individual iterations its value changes, so we use the notation $s^k \in \mathbb{R}^n$. We have already noted that denoting of vectors using boldface is omitted at present and even in this case we know from the context that $s^k \in \mathbb{R}^n$ is a vector [4].

Direction is usually not a problem to calculate, with the choice of step it is different. To make a new approximate solution x^{k+1} "better" than the old solution x^k, the step λ_k must be neither too small (short) nor too large (long).The optimal step length will correspond to the option λ_k, which will minimize the function on the line $x^k + \lambda s^k$. The optimal step is defined as the optimal solution of the following one-dimensional problem

$$\min\left\{ \varphi(\lambda) := F\left(x^k + \lambda s^k\right) \;\middle|\; \lambda \in \mathbb{R}\right\}. \tag{3.2}$$

Minimization problem (3.2) is an *unconstrained optimization problem of a single variable,* of the function $\varphi(\lambda)$.

Methods of minimising a function of n-variables $F(x)$, which use the step selection according to (3.2) are called ***optimal step methods*** (or ***perfect step methods***).

Selection of an optimal step in each iteration has an overall effect on convergence of a respective multidimensional optimization algorithm (it improves convergence rate and theoretically guarantees convergence of a relevant method). Therefore, it is desirable to implement at least an approximate solution of the problem (3.2) in each iteration (3.1). From the above it is clear that effective solution of the problem (3.2) will accelerate the whole iterative process of solving the original n-dimensional unconstrained optimization problem.

Therefore, it is important to have mastered *numerical methods for solving one-dimensional optimization problems* [2, 5, 6]. These include *comparative methods* that are used if derivative of an objective function is not known (or derivative does not exist). These methods can also be used for cases where we do not know a function explicitly, but its value can be measured at any point. A well-known and frequently used method is the *golden section method. Methods of the first and second kind* are suitable for functions for which derivative can be determined.

Since nonlinear programming is a superset of convex and linear programming, we will discuss these methods in terms of nonlinear programming methods.

3.1.1 One-Dimensional Minimization Methods

Specific position of unconstrained optimization problem of a single variable function is caused by the fact that it occurs as a separate subproblem in most optimization problems with multiple variables as it is used to determine an optimal step length in each iteration. Therefore, for successful optimization of multidimensional problems, knowledge of *numerical methods for finding a minimum of a single variable function* is a prerequisite.

Let us recall its formulation:

$$Min\{ f(x) \mid x \in \mathbb{R}\}, \tag{3.3}$$

or for some $a, b \in \mathbb{R}$ as

$$Min\{ f(x) \mid a \leq x \leq b\}. \tag{3.4}$$

The strategy for solving one-dimensional minimization problems is to convert *unconstrained optimisation problem* (3.3) to *constrained optimisation problem* (3.4), which in this situation is paradoxically classified as qualitatively simpler as a minimum of a function $f(x)$ is sought on a finite interval $[a, b]$. Numerical solution of *unconstrained (free) optimization problem) of a single variable function* takes place in two phases:

Phase I—***separating a minimum*** into a finite $[a, b]$, i.e., to reduce (convert) the problem (3.3) to the problem (3.4).
Phase II—solving the problem (3.4) and specify a solution.

Without certain simplifying assumptions, the problem (3.3) or (3.4) is practically unsolvable [4]. Therefore, a minimized function $f(x)$ will be assumed to be *strictly unimodal function* $f(x)$, i.e., it has only one extremum (for us it is a minimum) in a considered area of search.

Literature introduces other names for the phase I, e.g., "*determination of initial interval*", or "*uncertainty interval search*", [1, 6–9]. In each case, the objective is to determine a finite interval $[a, b]$, that contains a minimum x_0 of a function $f(x)$.

Separation of a minimum (1st phase of solving the one-dimensional problem)

- *First order method*
- *Zero order method*

First order method—uses a connection between the first derivative and a monotonicity of a function. If at the selected initial point x_1 it holds $f'(x_1) < 0$, then a function at this point is decreasing and we look for a minimum *to the right of the initial point* through an increasing sequence of points $\{x_k\}$. At these points x_k we gradually count derivatives $f'_k := f'(x_k)$ until a derivative change to a positive sign. If $f'_k < 0 < f'_{k+1}$, then a minimum x_0 must lie between the points x_k, x_{k+1}. We have determined the *boundaries* of the sought *uncertainty interval*: $a = x_k, b = x_{k+1}$ and $f'(a) \cdot f'(b) < 0$ is valid i.e., it is the so-called termination condition. We would proceed analogously even if we had a positive sign of derivation at the beginning at the initial point x_1 (i.e., $f'(x_1) > 0$, a function would be *increasing* at this point), we would only look for a minimum x_0 through the sequence of points $\{x_k\}$ to the left of the initial point x_1.

Zero order method—two initial points $x_1 < x_2$ are selected and a function value $f_1 := f(x_1)$ and $f_2 := f(x_2)$ are calculated. These two functional values determine the difference of a function $f_{12} := \frac{f_2 - f_1}{x_2 - x_1}$, which approximates a respective derivative, so we proceed as in the previous case. If $f_{12} < 0$, then a minimum x_0 lies to the right of the initial point x_1; if $f_{12} > 0$, then we look for the minimum to the left of the initial point.

The search for the initial interval $[a, b]$ is terminated if termination condition is met i.e., that in endpoints of an interval $[a, b]$ a derivative (or difference) of a function has a different sign. The above procedures include the idea of a given method. For this method to become an algorithm, rules must be laid down for generating corresponding monotonic sequences of points x_k, in general the distance between points is doubled. When algorithmizing this procedure, the *signum* function is used to shorten the notation.

Classification of methods for approximate solution of one-dimensional optimization (for phase II)

Approximate solution of the problem (3.4) can be understood in two ways:

(1) as a problem of determining a sufficiently fine uncertainty interval $[c, d]$ containing a sought minimum x_0, or
(2) as a problem of determining a point $\overline{x} \in [a, b]$ from a sufficiently small area around a searched point x_0.

In terms of the above approaches [1, 4] to the approximate solution of the problem (3.4), we can divide the relevant numerical methods into:

(i) *methods of interval approximation of a minimum* (*direct methods*),
(ii) *methods of point approximation of a minimum* (*interpolation methods*).

Within one-dimensional optimization methods *(i)* and *(ii)* we have another division in terms of information used (derivation order) regarding a minimized function $f(x)$:

- *zero order methods* that use only functional values f, (comparative methods);
- *first order methods*, which also use values of the first derivative f' of a function;
- *second order methods*, which also use values of the second derivative f'' of a function.

Zero order methods are universal, but they converge more slowly. Although first order methods require the existence of the first derivative, they are characterized by a better convergence rate. Second order methods require the existence of the second derivative and their local convergence is the best, but the volume of calculations is usually very large. For example, if we solve a problem of the optimal step (3.2), then the calculation of the first derivative f' requires calculation of n the partial derivatives of a function F (calculation of the gradient of the original function of n-variables), and calculation of the second derivative f'' requires calculation of $n(n+1)/2$ the second partial derivatives of a function F(calculation of the whole Hessian matrix of the original function of n-variables) [4, 10].

(i) ***Methods of interval approximation of a minimum (direct methods)*** are iterative methods for solving the problem (3.4), the principle of which is analogous to that of the phase I, i.e., it is based on a direct comparison of functional values of a minimized function at different points. The result of each iteration is a reduction in an uncertainty interval, and the *convergence rate does not depend on the form of a minimized function.* In general, the convergence of these methods is slow.

First order interval approximation methods include:

- ***Bisection method*** (also called the *Bolzano method*), is a sequential method, i.e., a *method of sequential experiments*. Calculation of one functional value f_k or calculation of one derivative value f_k' is called the one experiment. In the middle $c = \frac{a+b}{2}$ of an uncertainty interval $[a, b]$ we calculate a derivative $f'(c)$. If $f'(c) < 0$, then a new uncertainty interval containing a minimum x_0 will be an interval $[c, b]$ (taking into account that $f'(a) \cdot f'(b) < 0$ and $f'(a) < 0$). We can denote the new uncertainty interval $[a_1, b_1]$ and repeat the whole procedure. After k-th iteration for the length d_k of the uncertainty interval we obtain:

$$d_k = (b_k - a_k) = \frac{b-a}{2^k}.$$

  In these methods, we decide on the $(k+1)$-th experiment only after the evaluation of the k-th experiment (we decide on the position of a point x_{k+1}). In the bisection method, a sign of derivation f_k' is evaluated.
- ***Method of simultaneous experiments***—an equidistant network of points is selected at an interval $[a, b]$ which a priori decides on a position of evaluated points (evaluation is performed using f_k' and unimodality of a function $f(x)$ is used). Compared to the bisection method, this is a less efficient method because after n experiments the original interval reduces only $n+1$-times while bisection reduces up 2^n-times. Its application is important in expensive laboratory measurements of some short-term phenomenon.

Methods of zero-order interval approximation include:

- *Dichotomous method*
- *Method of simultaneous zero-order experiments*
- *Fibonacci method*
- *Golden section method*

Of all the methods of interval approximation, the most effective and the most frequently used is the ***golden section method***. We will describe this method separately, then we will present the principle and overview of interpolation methods for minimizing a function of a single variable.

Golden Section Method

This method has been widely applied in more complex algorithms for solving multidimensional optimization problems to find an optimal step length or an optimal value of a scalar parameter. The main idea of the *golden section method* (Table 3.2) is the application of previously used points in further decisions.

If an interval $[a, b]$ in the problem (2.16) is reduced by two experiments in points $c_1 < c_2$ on the interval $[a, c_2]$ (or on the interval $[c_1, b]$), then in the new interval the point c_1 (and the point c_2) will be with the already calculated value $f(c_1)$ and $f(c_2)$, respectively. This inherited value will save one experiment in the next iteration in

Table 3.2 Golden section algorithm

Input	Unimodal function $f(x)$ defined on $[a, b]$, Required accuracy $\varepsilon > 0$
Calculation	(0) $\varphi = \frac{\sqrt{5}+1}{2}$, $z_1 = 2 - \varphi$, $z_2 = \varphi - 1$ (1) $c_1 = a + z_1(b - a)$, $f_1 = f(c_1)$ $c_2 = a + z_2(b - a)$, $f_2 = f(c_2)$ (2) If $f_1 < f_2$, then go to (4) (3) The case $f_1 \geq f_2$ Let us put $a = c_1$, $c_1 = c_2$, $f_1 = f_2$ If $b - a < \varepsilon$ then go to (5), otherwise $c_2 = a + z_2(b - a)$ $f_2 = f(c_2)$, go to (2) (4) The case $f_1 < f_2$ Let us put $b = c_2$, $c_2 = c_1$, $f_2 = f_1$ If $b - a < \varepsilon$ then go to (5), otherwise $c_1 = a + z_1(b - a)$ $f_1 = f(c_1)$, go to (2) (5) $x_0 = c_1$, $f_0 = f_1$, end.

which a functional value is calculated at only one (new) point. Using this fact consistently, zero-order methods behave almost as effectively as the bisection method. In view of the above, the question naturally arises about the possibilities of optimal placement of experiments in each iteration The golden section method makes this possible by cleverly selecting two dividing points by dividing the original interval into three unequal parts (creating two overlapping intervals of equal length) with the dividing point c_1 (or c_2) dividing the original interval in the so-called *golden ratio*.

We say that a point divides an interval in the *golden ratio* if the two newly formed subintervals have the property that the ratio of the length of the whole interval to the length of the longer subinterval is the same as the ratio of the length of the longer subinterval to the length of the shorter subinterval [11]. The golden section method uses this property and thus maintains the similarity of interval divisions in individual iterations, which can be expressed for the division points c_1, c_2 by the following ratio

$$\frac{b-a}{c_2-a} = \frac{c_2-a}{c_1-a} =: \varphi$$

Let us suppose that an initial interval $[a, b]$ is a unit length or we can consider it as a unit interval $[0; 1]$, thus the relevant considerations on the golden section method will be simplified. We will divide the interval $[0; 1]$ by two experiments placed at points $z_1 < z_2$ symmetrically according to the centre, i.e., $z_1 = 1 - z_2$ on three unequal parts so that it holds $\frac{1}{z_2} = \frac{z_2}{z_1} =: \varphi$. After modification, we obtain a quadratic equation, the solution of which is the values $(-1 \pm \sqrt{5})/2$, of which only one belongs to the interval $[0; 1]$ and that is $(\sqrt{5} - 1)/2 \doteq 0,61803 = \tau$. We can express

$$z_2 = (\sqrt{5} - 1)/2 = \varphi - 1 = 1/\varphi = 0.618034\ldots,$$

$$z_1 = (3 - \sqrt{5})/2 = 2 - \varphi = 0.381966\ldots,$$

where $\varphi = 1.618034\ldots$, the quantity is also defined in the analysis of the Fibonacci method. The number $\tau = (\sqrt{5} - 1)/2 \approx 0,61803$ is a number known as the *golden ratio* [4, 5, 11–13].

Speed of length reduction d_n of uncertainty interval at n-th experiment is given by the relationship

$$d_n = \frac{b - a}{\varphi^{\,n-1}}$$

Although the *golden section method* is 17% worse than the Fibonacci optimal method, it does not suffer from the shortcomings of computational cumbersomeness, so it is more efficient, and it is preferred in practice.

The golden ratio (ratio for dividing lengths) has been an object of interest for more than 2400 years. It attracted not only mathematicians, naturalists, but also artists [4, 11, 12]. The first mention of this special relationship is attributed to the Pythagoreans, it appears in geometry, in architecture (Greek temple of the Parthenon), in art (Leonardo da Vinci and his proportions of the human body). It also occurs in living and inanimate nature (distances of branches and leaves, geometric ratio of some crystals). The *golden ratio* is often denoted φ, according to the Greek sculptor Feidias. In our text, we have denoted it as τ.

(ii) ***Methods of point approximation of a minimum (interpolation methods)*** are based on interpolation of a function $f(x)$ by some function funkciou $\phi(x)$ a minimum of which $\overline{x}$ can be expressed in a simple formula. Therefore, these methods are also called interpolation methods. It can be shown that Taylor expansion provides the best polynomial approximation of a given function around a given (working) point, i.e., that every other polynomial of the same degree gives a greater deviation from a given function. The Taylor formula can also be applied to a function of several variables, then the objective function f can be expressed by means of the Taylor series at a point $\boldsymbol{x}$ in the following form:

$$f(\boldsymbol{x} + \boldsymbol{p}) = f(\boldsymbol{x}) + g(\boldsymbol{x})^T \boldsymbol{p} + \frac{1}{2}\boldsymbol{p}^T \boldsymbol{H}(\boldsymbol{x})\boldsymbol{p} + \cdots, \tag{3.5}$$

where $\boldsymbol{p}$ expresses a change vector, $g(\boldsymbol{x})$ is a gradient and $\boldsymbol{H}(\boldsymbol{x})$ is Hessian function f.

One of the possibilities of dividing optimization methods is their classification according to the number of terms of Taylor series used in the interpolation of the objective function f.

Direct methods of finding an extremum use only values of a function f, in *gradient methods* there is used a *linear approximation* of a function f in the vicinity of the stationary point and *quadratic approximation* of the function f v blízkom okolí stacionárneho bodu a *kvadratickú aproximáciu* funkcie f uses *Newton's method* and some of its modifications [2, 14, 15].

General scheme of interpolation methods for one-dimensional optimization:

Input: given *interpolation nodes* x_i $(i = 1, 2, \ldots, k)$, $k \leq r$ and corresponding values $f_i = f(x_i)$, or $f_i' = f'(x_i)$. In addition, we have a given class of functions that depends on r parameters (e.g., the class of quadratic functions has three parameters).

1. *Construction of interpolation function* $\phi(x)$ is based on calculation of values of the relevant parameters so that they apply in interpolation nodes $\phi'(x_i) = f'$, or $\phi'(x_i) = f'$.
2. *Calculation of a minimum* $\overline{x}$ of interpolation function $\phi(x)$, which approximates the required minimum $\hat{x}$ of a function $f(x)$.
3. If $\overline{x}$ is good enough approximation of the required minimum $\hat{x}$, then we finish calculation. Otherwise, we replace by point $\overline{x}$ the *worst* of the used interpolation nodes. We calculate $\overline{f} = f(\overline{x})$ or also $\overline{f}' = f'(\overline{x})$ and we return to step 1.

In practice, as *the worst* of the used interpolation nodes is considered a such node that is furthest from the approximation of a minimum. A *sufficiently good approximation* of a minimum can be considered such a point which its distance from the approximation obtained in the previous iteration is less than the predetermined tolerance constant [4].

We obtain specific *interpolation methods* depending on:

- options for the type of interpolation function $\phi(x)$, quadratic parabola, cubic parabola, quadratic spline, etc., can be used,
- the method of interpolation, i.e., how many interpolation nodes we have available and what information we use about a minimized function $f(x)$ in individual nodes (functional value or derivative, or both functional value and derivative).

Quadratic interpolation of the minimum—for interpolation function we choose a quadratic parabola, which can be shifted. If the shift of the argument x is expressed by a parameter z, then the parabola can be written in the following form:

$$\phi(x) = a(x - z)^2 + b(x - z) + c \tag{3.6}$$

The shift parameter z simplifies interpolation equations for calculating unknown parameters a, b, c. If $a > 0$, then a minimum $\overline{x}$ of the function (3.6) will be obtained based on meeting the necessary condition for the existence of an extremum, i.e., from the equation $\phi'(x) = 0$ in the following form:

$$\overline{x} = z - \frac{b}{2a} \tag{3.7}$$

In the relation (3.7) only two parameters a, b appear, because the additive constant c drops out by derivation. It follows from the above that in general we need *three information units* (functional values or derivatives), but there are special cases where two are enough. Methods using quadratic interpolation vary in the number of interpolation nodes.

The case of one interpolation node: we approximate a function $f(x)$ around a node x_1 by the second-degree Taylor polynomial:

$$f(x_1) + f'(x_1)(x - x_1) + \frac{1}{2} f''(x_1)(x - x_1)^2$$

If we compare this polynomial with the function (2.18), then we see that $z = x_1$, $2a = f_1''$, $b = f_1'$, $c = f_1$ and substituting into the (2.18) we get the following formula

$$\overline{x} = x_1 - \frac{f_1'}{f_1''} \tag{3.8}$$

If a found minimum point $\overline{x}$ is chosen as a new interpolation node, the whole procedure is repeated and then we get the so-called second order Newton's method (it uses the knowledge of the first and second derivatives) to calculate a minimum of a function of a single variable. It is also known as the *tangent method,* or the ***Newton–Raphson method*** based on the solution of a nonlinear equation $f'(x) = 0$ similar to bisection. If we choose an initial point x_1, then a sequence of approximations is defined by the following relation:

$$x_{k+1} = x_k - \frac{f'(x_k)}{f''(x_k)} \text{ pre } k = 1, 2, \ldots \tag{3.9}$$

The advantage of the *Newton–Raphson method* lies in fast convergence, the disadvantage is that it is necessary to choose a node (initial point) x_1 close enough to a point of a search minimum $\hat{x}$.

The case of two interpolation nodes: here are possible three ways of entering interpolation information in given nodes (assuming we consider only functional values and derivatives in nodes $x_1 < x_2$).

In the first method, quantities $f_1' < 0 < f_2'$ are given. Using suitable substitutions and modifications [6, 7, 16], we get the relation for calculating a minimum $\overline{x}$ of an interpolation function $\phi(x)$ in the following form

$$\overline{x} = x_1 - (x_2 - x_1)\frac{f_1'}{f_2' - f_1'} = x_2 - (x_2 - x_1)\frac{f_2'}{f_2' - f_1'}, \tag{3.10}$$

which is an intersection of a secant intersected by endpoints (nodes) with the axis x. This intersection makes it possible to narrow an uncertainty interval and repeat the procedure in the next step, so that the relation (3.10) is the basis for the so-called *secant method* and its modification—the ***regula falsi method***, where from geometric approach or from the *quadratic two-point interpolation of the objective function* $f(x)$ [8, 10, 17–19] we get the iterative relationship

$$x_{k+1} = x_k - (x_k - x_{k-1})\frac{f'(x_k)}{f'(x_k) - f'(x_{k-1})}. \qquad (3.11)$$

In the second method of entering interpolation information, quantities $f_1' < 0, f_1, f_2$; are given; in the third method $f_2' > 0, f_1, f_2$, are given. We can find more in [16] where the case of three nodes is also presented.

Cubic interpolation of a minimum—for z interpolation function we choose a cubic parabola, which can be written in the form with a shift argument

$$\varphi(x) = a(x-z)^3 + b(x-z)^2 + c(x-z) + d, \qquad (3.12)$$

where a shift argument x is expressed by a parameter z, which is appropriately chosen. To determine four parameters (a, b, c, d) we need four conditions, so we use interpolation conditions. If we investigate the first and second derivatives of a functiosn $\varphi(x)$ (3.12), o we get the following statement:

Statement 3.1.1 *If* $a \neq 0$, $b^2 \geq 3ac$, *then the function* (3.12) *has a (local) minimum at a point*

$$\overline{x} = z + \frac{-b + \sqrt{b^2 - 3ac}}{3a} = z + \frac{-c}{b + \sqrt{b^2 - 3ac}}. \qquad (3.13)$$

If $a = 0$, then $b > 0$ must be valid. In the expression (3.13) the parameter d does not occur. This means that while in general for interpolation by cubic parabola (3.12) we need to enter four information units, there are cases where three information units are sufficient: if we have only derivatives entered in interpolation nodes. In practice, *cubic interpolation* of a minimum *with two interpolation nodes* $x_1 < x_2$, is used very often, when the values $f_1' < 0 < f_2', f_1, f_2$ are given. This is also the case of the following method.

Davidon's cubic interpolation method—is one of the most frequently used interpolation methods of one-dimensional optimization. It is very often used in various optimization programs when finding an optimal step length in solving multidimensional optimization problems. This method is computationally efficient.

If in the relation (3.12) we choose $z = (x_1 + x_2)/2$ and introduce auxiliary denotations $h = (x_2 - x_1)/2$, $f_{12} = (f_2 - f_1)/(x_2 - x_1)$, thus, by using interpolation conditions and breaking them down, we obtain a system of four equations with four unknowns (a, b, c, d). However, we can reduce this by suitable modifications to a system of two equations of two unknowns. Having solved them, we obtain values a, b, c. For more detail, see [4, 6, 8]. If we use *Statement 3.1.1* and substitute these values into the formula (3.13), we obtain the resulting relation for approximation of an extreme on an interval $[x_1, x_2]$ in the following form

$$\overline{x} = x_1 + (x_2 - x_1)\frac{W + Z - f_1'}{f_2' - f_1' + 2W}, \tag{3.14}$$

where for auxiliary variables Z, W the following is valid

$$\begin{aligned} Z &= f_1' + f_2' - 3f_{12} \\ W &= \sqrt{Z^2 - f_1' f_2'} \end{aligned} \tag{3.15}$$

Another point is another approximation of an extremum (a minimum) of an investigated function is calculated analogously on a narrow interval $[x_1, \overline{x}]$ if $f'(\overline{x}) > 0$, or $[\overline{x}, x_2]$ if $f'(\overline{x}) < 0$. A termination condition is usually specified by the prescribed tolerance.

Davidon's cubic interpolation method is numerically efficient in terms of a required volume of calculations in each step as well as the convergence rate. It is part of more complex optimization algorithms. The formula (3.14) was first published in 1963 by Fletcher and Powell [16], in which they developed the basics of the so-called quasi-Newton methods to minimize a function of n-variables.

Quadratic spline interpolation—A quadratic spline is chosen as an interpolation function and the so-called glued point is defined. For more detail, see [1, 4].

3.1.2 Methods for Minimization a Function of n Variables

The subject of our interest regarding optimization problems in engineering activities are *numerical methods for multidimensional minimization.*

Classification of methods for approximate solution of multidimensional optimization tasks

Classification of these methods is based on the classification of multidimensional optimization problems, i.e., the problems (U1)–(U4), which can be recalled from the Sect. 2.3.4. Accordingly, we distinguish:

(i) *methods for solving minimization problems without constraints, i.e. methods for solving unconstrained optimization problems* (U1) *(also so-called free optimisation problems),*
(ii) *methods for solving minimization problems with constraints, i.e., methods for solving* (U2)–(U4) *problems.*

Multidimensional optimization methods for OP without constraints, i.e., methods presented in the category *(i)* are further subdivided in terms of information used (derivation order) on a minimized function $f(\boldsymbol{x})$, $\boldsymbol{x} \in \mathbb{R}^n$. We distinguish the following methods:

- *zero order methods* (*direct methods*) that use only functional values of f; these are mainly methods using a pattern ("*pattern search*"), e.g., the Hooke–Jeeves method, the ***Nelder-Mead method***;
- *first order methods* that also use values of the first derivative f' of a function; these include *classical methods*: gradient method, the ***Cauchy's steepest descent method***, ***Newton's method***, *modern methods:* the *Fletcher-Reeves conjugate gradient method*, ***quasi- Newton's methods, BFGS method*** Fletchera-Reevesa, ***kvázinewtonovské metódy***, ***metóda BFGS***;
- *second order methods* that also use values of the second derivative f'' of a function (the Newton–Raphson method, the Levenberg–Marquardt method).

Multidimensional optimization methods for OP with constraints, i.e., methods presented in the category *(ii)* are further subdivided in terms of *constraint type*. The most used methods are as follows:

- *methods for solving minimization problems with constraints in the form of equations,* i.e. the classical problem of extremum subject to constrains (methods for solving (U2) problem); for example the Lagrange multipliers technique, ***reduced gradient method***, Newton's method using reduced Hessian matrix, penalty method;
- *methods for solving minimization problems with constraints in the form of inequalities or in the form of equations and inequalities,* (methods for solving (U3), (U4) problems); for example: method of possible directions, *barrier functions and their use*, penalty method, complex method.

Methods marked in bold will be presented in more detail in Chap. 4, which will be devoted to selected optimization methods with respect to application in optimization of technological processes.

Now we will briefly discuss the *classification of methods for solving unconstrained (free) optimisation problem* from other points of view. Free optimisation problem is the optimization problem of the following type:

$$Min\{ f(x) \mid x \in \mathbb{R}^n \} \quad \text{(U1)}$$

Almost all the methods of solving the problem (U1) are iterative, i.e., they generate a sequence of points converging to an optimal solution of the problem (U1) or converging at least to a local minimum point of a function $f(x)$, $x \in \mathbb{R}^n$. An essential feature of each iteration algorithm is the ability to generate at each step a better approximation of a minimum than the previous one, which expresses an *objective rate of progress*. For problems (U1), the natural measure of iteration success is a decrease in a functional value of an objective function, i.e., the fulfilment of the relationship

$$f\left(x^{k+1}\right) < f\left(x^{k}\right), \quad (k = 0, 1, 2, \ldots) \tag{3.16}$$

Gradient descent method is an iterative method for the solution of the (P1), which has the property of the (3.16).

Autonomous iterative method (one-step or simple) is a method that has the following form

$$x^{k+1} = \mathcal{I}\left(x^{k}\right), \quad (k = 0, 1, 2, \ldots) \tag{3.17}$$

where $\mathcal{I} : \mathbb{R}^n \to \mathbb{R}^n$ mapping characterizing given iterative method. We see that autonomous iteration methods depend only on the immediately preceding iteration x^k and do not depend on older iterations. They are classical methods such as the *Cauchy steepest descent method* and the *Newton's method.*

Non-autonomous iterative method (multi-step or memory method) is a method of the following form

$$x_{k+1} = \mathcal{I}\left(x^{k}, x^{k-1}, \ldots, x^{k-r}\right), \quad (k = r, r+1, r+2, \ldots) \tag{3.18}$$

Non-autonomous methods have a more complicated representation of $\mathcal{I}$, however, compared to autonomous methods, they use the available information about an objective function more rationally, so they are better, more efficient, and currently have a wider application. The *combined conjugate-gradient* or *quasi-Newton method* is an example.

The iterative formulas (3.17), (3.18) are usually written in the form of an additive correction of the previous iteration as follows

$$x^{k+1} = x^{k} + p^{k}, \quad (k = 0, 1, 2, \ldots), \tag{3.19}$$

or as

$$x^{k+1} = x^{k} + \lambda_k s^{k}, \quad (k = 0, 1, 2, \ldots) \tag{3.20}$$

where p^k, s^k are non-zero n-dimensional vectors and λ_k is a non-zero scalar, where

$$\lambda_k s^{k} = p^{k} = x^{k+1} - x^{k} \tag{3.21}$$

is the so-called *correction vector* (shift vector), which is a suitable λ_k multiple of some direction s^k. The correction vector (3.21) is determined in specific methods in two phases according to two different schemes:

I. *(Point, Direction, Step)* In the first phase, a *descent direction* s^k is determined at a point x^k and then, in the second phase, a step length λ_k is determined. The new approximation x^{k+1} is then defined by the relation (3.19).

II. *(Point, Step, Direction)* In the first phase, a step length Δ_k is determined at a point x^k and then, in the second phase, a *direction* p^k is determined so that $\| p^k \|_2 \leq \Delta_k$. The new approximation x^{k+1} is then defined by the relation (3.20).

Step length control methods—are based on the Scheme I, forming most of the iterative methods. They can be further classified according to the direction selection method or according to the step selection method. They differ from each other mainly by the selection of direction s^k (there is an unlimited number of selection strategies). The choice of a step λ_k offers only two fundamentally different strategies by application of which we recognize either methods with a *constant step length* or methods with an *optimal step length.* We have already mentioned the latter in the one-dimensional optimization, where we find an optimal step length λ_k by solving the problem (2.14) by finding a minimum of a function $\varphi(\lambda) := f(x^k + \lambda s^k)$, i.e.,

$$Min\{\varphi(\lambda) := f(x^k + \lambda s^k) \mid \lambda \in \mathbb{R}\} \tag{3.22}$$

Direction control methods—are based on the Scheme II and they are known as the *constrained step method.* They did not appear until the 1960s.

Selection of directions s^k in the relation (3.21) can be realized according to one of the following principles:

(i) In each iteration, a direction s^k is selected randomly or according to a certain probabilistic law (the so-called ***stochastic optimization algorithms***).
(ii) A system of n linearly independent directions is selected in advance. Then, in individual iterations, directions (usually cyclically) are gradually selected from this system.
(iii) Direction s^k depends on a point x^k as well as on a minimized function. In case of non-autonomous methods, it also depends on the previous points and in the steepest descent method it holds $s^k = -\nabla f(x^k)$. These methods involve transformation of a gradient direction $\nabla f(x^k)$, so these methods are known as gradient methods.

In the cases (i) and (ii), the length of a step may also be expressed by a negative number as a relevant direction need not be a descent direction. The situation is different in the case (iii), where a direction s^k is defined to be descent, i.e., that a value of an objective function of the problem (U1) decreases in this direction, which means that a step length λ_k will always be a positive number. In *interior point methods*, where a function on an open set M is minimized, a step length is selected so that the new iteration point remains at M. In terms of the Definitions 2.1.6a, 2.1.6b, we can introduce the following definition.

Definition 3.1.1 *A direction s^k will be called descent direction of a function $f(x)$ defined on $\mathbb{R}^n$ at a point $x^k \in \mathbb{R}^n$, if there exists a positive number $\overline{\lambda}$ so that for all $0 < \lambda < \overline{\lambda}$ the following is valid:*

$$f(x^k + \lambda s^k) < f(x^k) \tag{3.23}$$

Using elementary properties of a derivative of a function, the following auxiliary statements can be proved based on the mentioned definition:

Lemma 3.1.2 *Let a function $f(x)$ defined on $\mathbb{R}^n$ have continuous first partial derivatives. Then a direction $s^k \in \mathbb{R}^n$ is descent at a point $x^k \in \mathbb{R}^n$ if the following is valid*

$$\nabla f\left(x^k\right)^T s^k < 0 \tag{3.24}$$

Consequence 3.1.2 *The so-called* ***Cauchy direction***

$$s^k = -\nabla f\left(x^k\right) \quad \textit{is descent.} \tag{3.25}$$

Lemma 3.1.3 *Let a function $f(x)$ defined on $\mathbb{R}^n$ have continuous first partial derivatives and let H_k be some positive definite matrix. Then a direction*

$$s^k = -H_k \nabla f\left(x^k\right) \quad \textit{is descent.} \tag{3.26}$$

Consequence 3.1.3 *Let a function $f(x)$ defined on $\mathbb{R}^n$ have continuous second partial derivatives and let at a point $x^k \in \mathbb{R}^n$ its Hessian matrix of the second partial derivatives be positive definite. Then the so-called* ***Newton direction***

$$s^k = -\left[\nabla^2 f\left(x^k\right)\right]^{-1} \nabla f\left(x^k\right) \quad \textit{is descent.} \tag{3.27}$$

The Cauchy direction is a direction of the steepest descent because a gradient $\nabla f\left(x^k\right)$ represents an outer normal at a point to the graph of a function. However, this feature does not yet guarantee good convergence even with an optimal selection of a step length.

Since even the combination of the locally best direction with the best step may not yet give a good result, the requirement to calculate an optimal step length as the most challenging part of the iteration has begun to lose its relevance. Therefore, in today's efficient optimization algorithms, we pass to the requirement of an *approximately optimal step*. However, in order not to endanger convergence of relevant methods by this concession, it has proved expedient to introduce certain safety limits for a step length. Possible ways of determining such safety limits applicable in practice have been shown by Goldstein (*upper limit of step length*) or Wolf and Powel (*lower limit of step length*) and known convergence theorems (Wolfe, 1971), (Dennis-Moré, 1974). See more in [4, 8, 20, 21].

Zero-order methods of multidimensional minimization without constraints are universal, but relatively slowly converging. They are used if an objective function is not differentiable.

The first-order methods used for solving (U1), which require the existence of an objective function gradient, have better and more satisfactory convergence than zero-order methods.

Second-order methods (e.g. *Newton's method*), requiring the existence of the Hessian matrix of second partial derivatives, have very good convergence locally, but too large computation volume in one iteration until recently limited the use of these methods to small-scale problems. However, under the influence of modern methods of an initial point there was a renaissance of the Newton method. The special structure of the Hessian matrix is used, and special techniques have been developed to calculate the so-called Newton directions even in the case of large-scale optimization problems. The special structure of the Hessian matrix is also used for a logarithmic barrier function of the linear programming problem.

Selected methods of one-dimensional optimization (golden section method, bisection method, etc.) are presented in this chapter.

Selected methods and algorithms of multidimensional optimization for free boundary optimization problems and especially for constraint optimization problems will be presented in Chap. 4. Firstly, we will present a general view of the evaluation of a correct functioning of individual optimization methods and justify the development of the so-called stochastic optimization algorithms with respect to their use in practice.

3.2 Testing of Optimization Algorithms

The correctness of the functioning of optimization methods of finding an extremum began to be verified using special, the so-called test functions, for which we know the extreme value. The specialty of the test functions lies in the fact that they cause various problems in finding an extremum and thus allow to compare the efficiency of optimization algorithms. In addition to verifying the position and extreme value, they can help us to describe the characteristics of the optimization algorithm (convergence rate, accuracy, and robustness of methods) [22, 12, 14]. Test functions differ in the number of variables, the number of stationary points and their structure. In [12, 20, 21, 23], we find a comparison of different methods and the authors present the results of numerical experiments. A clear summary of test functions is shown in [8], we will mention only some of them. The most famous are the following functions.

The *Rosenbrock's function*, ("banana" function), belongs to the unimodal functions (Fig. 3.1). This is a test function in the following form

$$f(\boldsymbol{x}) = (x_1 - 1)^2 + 100\left(x_2 - x_1^2\right)^2 \tag{3.28}$$

It is obvious that it has a single minimum at a point $\boldsymbol{x}^* = (1;\ 1)^T$.

Walukiewicz generalized this type of a function for functions of n variables

$$f(\boldsymbol{x}) = \sum_{i=1}^{n-1}\left[100\left(x_{i+1} - x_i^2\right)^2 + (x_i - 1)^2\right] \tag{3.29}$$

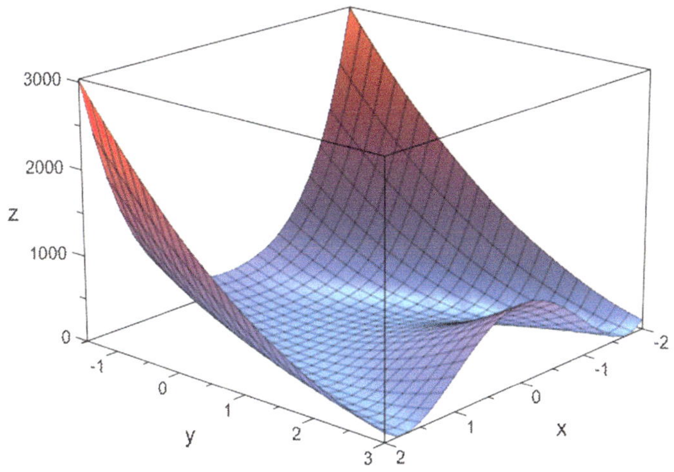

Fig. 3.1 Rosenbrock's function in $\mathbb{R}^3$, the only minimum is in $\boldsymbol{x}^* = (1;\ 1)^T$

which in the minimization problem $Min\{f(\boldsymbol{x})\ |\boldsymbol{x} \in \mathbb{R}^n\}$ acquire the minimum at a point $\boldsymbol{x}^* = (1, 1, \ldots, 1)^T$ and functional value $f(\boldsymbol{x}^*) = 0$.

Although the "valley" of a function is not difficult to find, convergence to a minimum is quite complicated.

White and Holst proposed three modifications of the Rosenbrock's function [12], *Himmelblau* proposed a test function

$$f(\boldsymbol{x}) = \left(x_1^2 + x_2 - 11\right)^2 + \left(x_1 + x_2^2 - 7\right)^2 \quad (3.30)$$

which in the minimization problem $Min\{f(\boldsymbol{x})\ |\boldsymbol{x} \in \mathbb{R}^n\}$ acquires four minima, among them at the point $\boldsymbol{x}^* = (3;\ 2)^T$, as seen in Fig. 3.2.

Other known test functions are from Powell and Fletcher, Eason, and Fenton. Poljak generalized the Box function. Part of the Wood function of the four variables is the Rosenbrock function, and its minimum is at a point $\boldsymbol{x}^* = (1; 1; 1; 1)^T$. The Powell function of four variables has a minimum at a point $\boldsymbol{x}^* = (0; 0; 0; 0)^T$. The shape of individual functions is presented in [12].

3.3 Stochastic Optimization Algorithms

These are very popular optimization algorithms in recent days. The term "stochastic" means that in an optimization process of finding an optimal solution from feasible region, a random element (random number) is implemented in a certain way. It is therefore not a deterministic method and a solution obtained using this approach is random. In many cases, it is not even possible to speak of convergence of such methods. Nevertheless, they have gained popularity and application in a wide range

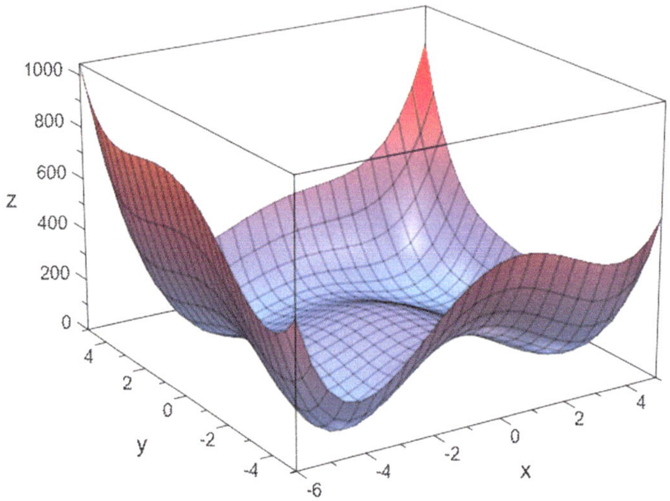

Fig. 3.2 Himmelblau function in $\mathbb{R}^3$

of technical and natural sciences. Although it is almost impossible to make a statement for a solution obtained in this way. Practice and many tests of these methods have shown their advantages mainly in speed and ability to handle time-consuming and computationally intensive solutions to optimization problems with complicated constraints and an objective function. The use of these methods in practice has shown that sometimes it is more appropriate to accept the solution achieved, which is in some sense sufficiently optimal (satisfactory for us) than to find a real optimum (if possible) at the cost of large time requirements.

These are relatively young methods, developed mainly in the second half of the 20th century because their development was closely connected with the development in the field of computer technology as their use is conditioned by sufficient computing capacity of computers. Their development has practically not stopped, and their new modifications and improvements are still being added, and therefore they are also used in engineering optimization. Such algorithms include evolutionary, genetic algorithms, inspired by evolutionary behaviour in biology and genetics [2, 9, 24–29].

In Chap. 4, we describe the principle of operation of selected methods of these types. We will focus on the ***Nelder-Mead method*** and simulated annealing. Using these algorithms, it is possible to solve problems with nonconvex purpose function and nonlinear constraints [1, 2, 10, 24, 26, 27, 30–32].

References

1. Rao, S. (2009). *Engineering optimization. Theory and practice* (4th ed., 830p.). John Wiley & Sons, Inc.
2. Boyd, S., & Vandenberghe, L. (2009). *Convex optimization* (701p.). Cambridge University Press. ISBN 978-0-521-83378-3.
3. Dupačová, J., & Lachout, P. (2011). *Úvod do optimalizace* (81p.). Matfyzpress. ISBN 978-80-7378-176-7.
4. Hamala, M., & Trnovská, M. (2012). *Nelineárne programovanie/Nonlinear programming* (339p.). Epos. ISBN 978-80-805-7986-9.
5. Brunovská, A. (1990) *Malá optimalizácia* (248p.). Alfa. ISBN 80-05-00770-1.
6. Rosinová, D., & Dúbravská, M. (2008). *Optimalizácia.* STU v Bratislave. 195 s. ISBN 978-80-227-2795-2.
7. Hudzovič, P. (1990). *Identifikácia a modelovanie/Identification and modeling* (2nd ed., 255p.). Slovenská vysoká škola technická v Bratislave. ISBN 80–227–0213–7.
8. Hudzovič, P. (2001). *Optimalizácia* (320p.). STU v Bratislave. ISBN 80–227–1598–0.
9. Schrijver, A. (2017). *A course in combinatorial optimization* (p. 221). University of Amsterdam.
10. Marčuk, G. I. (1987). *Metódy numerické matematiky* (528p.). Academia, nakladatelství ČSAV.
11. Yang, W.Y., Cao, W., Chung, T.S., Morris, J. *Applied Numerical Methods Using MATLAB*; John Wiley & Sons, Inc.: Hoboken, NJ, USA, 2005; p. 509.
12. Cao, H., Qian, X., & Zhou, Y. (2018). Large-scale structural optimization using metaheuristic algorithms with elitism and a filter strategy. *Structural and Multidisciplinary Optimization, 57*, 799–814. https://doi.org/10.1007/s00158-017-1784-3.
13. Čermák, L., & Hlavička, R. (2016). *Numerické metódy. Numerical methods* (110p.). CERM. ISBN 978-80-2145437-8.
14. Hliník, J. (2015). *Tvarová optimalizace difuzoru vodní turbíny* (55p.). Bachelor's thesis, VUT v Brně.
15. Machalová, J., & Netuka, H. (2013). *Numerické metody nepodmíněné optimalizace* (142p.). 1. vyd. Univerzita Palackého v Olomouci. ISBN 978–80–244–3403–2.
16. Hamala, M. (1972). *Nelineárne programovanie.* Alfa.
17. Ivan, J. (1989). *Matematika 2* (632p.). Alfa. ISBN 80–05–00114–2.
18. Jarník, V. (1976). *Diferenciální počet I.* Academia.
19. Rektorys, K. et al. (1981). *Přehled užité matematiky.* SNTL. 1140 s.
20. Реклейтис, Г., Рейвиндран, А., Рэгсдел, К. (1986) *Оптимизация в технике* 1, [Reklaitis, G. V., Ravindran, A., & Ragsdell, K. M. *Engineering optimization. Methods and applications*] (350p.). Mir, Moskva.
21. Реклейтис, Г., Рейвиндран, А., & Рэгсдел, К. (1986). *Оптимизация в технике* 2, [Reklaitis, G. V., Ravindran, A., & Ragsdell, K. M. *Engineering optimization. Methods and applications*] (320p.). Mir, Moskva.
22. Avriel, M., Rijckaert, M. J., & Wilde, D. J. (1973). *Optimization and design* (2nd ed., p. 489). Cambridge University Press, Prentice Hall.
23. Jablonský, J., Fiala, P., & Maňas, M. (1985). *Vícekriteriální optimalizace.* Praha, Czech Republic.
24. Ban, N., & Yamazaki, W. (2021). Efficient global optimization method via clustering/classification methods and exploration strategy. *Optimization and Engineering, 22*, 521–553. https://doi.org/10.1007/s11081-020-09529-4.
25. Chai, R., Savvaris, A., Tsourdos, A., Chai, S., Xia, Y. (2019) A review of optimization techniques in spacecraft flight trajectroy design. *Progress in Aerospace Sciences*, 109 (2019). Elsevier Ltd.
26. Khan, S., Asjad, M., & Ahmad, A. (2015). Review of modern optimization techniques. *International Journal of Engineering Research & Technology (IJERT), 4*(04). https://doi.org/10.17577/IJERTV4IS041129.

27. Locatelli, M. (2002). Simulated annealing algorithms for continuous global optimization. In P. M. Pardalos & H. E. Romeijn (Eds.), *Handbook of global optimization. Nonconvex optimization and its applications* (Vol. 62). Springer. https://doi.org/10.1007/978-1-4757-5362-2_6.
28. Yang, Z. (2021). On the step size selection in variance-reduced algorithm for nonconvex optimization. *Expert Systems with Applications*, *169*, 114336, 12p. https://doi.org/10.1016/j.eswa.2020.114336.
29. Zhou, D., Xu, P., & Gu, Q. (2018). Stochastic nested variance reduction for nonconvex optimization. In *International conference on neural information processing systems* (pp. 3925–3936). Curran Associates Inc.
30. Antoniou, A., & Lu, W. S. (2007). *Practical optimization. Algorithms and engineering applications* (675p.). Springer Science & Business Media LCC. ISBN-13: 978-0-387-71106-5.
31. Kirkpatrick, S., Gelatt, C. D., & Vecchi, M. P. (1983). Optimization by simulated annealing. *Science, 220*(4598), 671–680. https://doi.org/10.1126/science.220.4598.671
32. Taufer, I., Drábek, O., & Javůrek, M. (2010). Metoda simplexů—efektivní nástroj pro řešení optimalizačních úloh. *Řízení a automatizace, XX*(6).

Chapter 4
Selected Methods of Multidimensional Optimization

Rapid development of computer technology that has occurred in recent decades is closely linked to the development of computational algorithms capable of processing enormous amounts of data. In the past, many methods have been developed that can elegantly solve optimization problems and find an optimal solution, but these required strict fulfilment of prerequisites for an objective function and constraints very often. In addition, their effectiveness was hampered by the problem of large amounts of data because it was time consuming. Therefore, the question arose as to whether the algorithm was able to find a solution in polynomial time. In practice, moreover, it is sometimes necessary to obtain a solution in a relatively short time, so in some optimization problems the speed of calculation at the expense of the accuracy of the result, which in many cases was sufficient information, has begun to emphasize. This was the impetus for the development of modern optimization methods. Many of them can be classified as heuristic (discovery) algorithms. It is an algorithm that gradually searches (discovers) for better solutions. The advantage over explicit methods is the ease of input assumptions and the ability to cope with local extrema. The disadvantage is that in general, there is no evidence of their convergence to a global optimum. These methods include the *Nelder-Mead method* and other stochastic algorithms, e.g., *differential evolution method* and *simulated annealing method.*

In the next sections, we will present the selected methods for solving *nonlinear programming* problems (direct and gradient methods, quasi-Newton methods, BFGS) and *methods for linear programming,* which can be described as *classical optimization methods*, but their use is still justified in engineering practice. The *gradient methods* that use information about the first and second derivatives also include *interior point methods* are suitable for solving nonlinear programming and linear programming problem.

A. Vagaská et al., *Optimization Methods in Mathematical Modeling of Technological Processes*, Mathematical Engineering, https://doi.org/10.1007/978-3-031-35339-0_4

4.1 Selected Methods of Nonlinear Programming

The beginnings of the development of nonlinear programming date back to around 1951 when well-known Karush-Kuhn-Tacker publicised their optimality conditions (KKT). See more in [1–4]. Nonlinear programming problems, i.e., a general form of a problem (NP) as well as the special cases (U1)–(U4), were defined in Sect. 2.3.4, where we present the selected numerical methods for solving nonlinear programming problems. From *direct methods* that do not use information about derivation of an objective function, we present the Nelder-Mead method that includes genetic and stochastic algorithms. We introduce the method of simulated annealing (SA) as well as use the *gradient methods* of the *Cauchy steepest descent method,* the *Newton method* and the quasi-Newtonian *BFGS method* that seems to be the most effective of them.

4.1.1 Nelder-Mead Simplex Method

As stated in [3, 5–7], this minimization method based on finding an optimum using a simplex was proposed by Spendley, Hext and Himsworth in 1962. In the original variant, the authors used only regular *simplexes,* which are the simplest polyhedra in n-dimensional space that have $n + 1$ vertices. For the case $n = 1$, i.e., on a straight line $\mathbb{R}^1$, the line segment determined by two points is considered to be a simplex. For the case $n = 2$, i.e., in the plane $\mathbb{R}^2$, the regular simplex has 3 vertices, it is an equilateral triangle. For $n = 3$ it is a regular quadrilateral in the Euclidean space $\mathbb{R}^3$, etc. In 1965, British mathematicians Nelder and Mead modified the original method of regular simplexes so that the shape of the simplexes could be deformed. i.e., stretched or compressed as needed, allowing a better response to the current distribution of functional values at individual vertices of a simplex [5, 8–10].

The *Nelder-Mead method,* also known as the *simplex* method, or the *method* of *"flexible" simplex"* as presented in [3, 11], is a popular method of minimizing a function of several variables that does not use a derivation of an objective function. It is only a coincidence that the basic method of solving linear programming problems, the principle of which is to move along the edges of a polyhedron, is called the simplex method. The *Nelder-Mead simplex method* is different. It is one of the comparative methods, which are methods looking for a minimum of an objective function f by comparing its values at certain points in space $\mathbb{R}^n$. In the case of the *Nelder-Mead simplex method*, the selected points are simplex vertices [5, 11].

The main idea of one step of the method is simple: among the vertices $\boldsymbol{x}_0, \boldsymbol{x}_1, \ldots, \boldsymbol{x}_n$ of the simplex we choose the *worst* vertex $\boldsymbol{x}_w$, in which an objective function acquires the greatest value, and replace it with a better vertex $\hat{\boldsymbol{x}}$, in which a value of an objective function is smaller. We look for a vertex $\hat{\boldsymbol{x}}$ on a semi-line, which rises from the worst peak $\boldsymbol{x}_w$ and passes through the centre of gravity $\overline{\boldsymbol{x}}$ of remaining peaks, see Fig. 4.1 (for the sake of clarity, if $n = 2$, i.e., in the plane $\mathbb{R}^2$). We will denote $\boldsymbol{x}_b$, i.e., $\boldsymbol{x}_b$

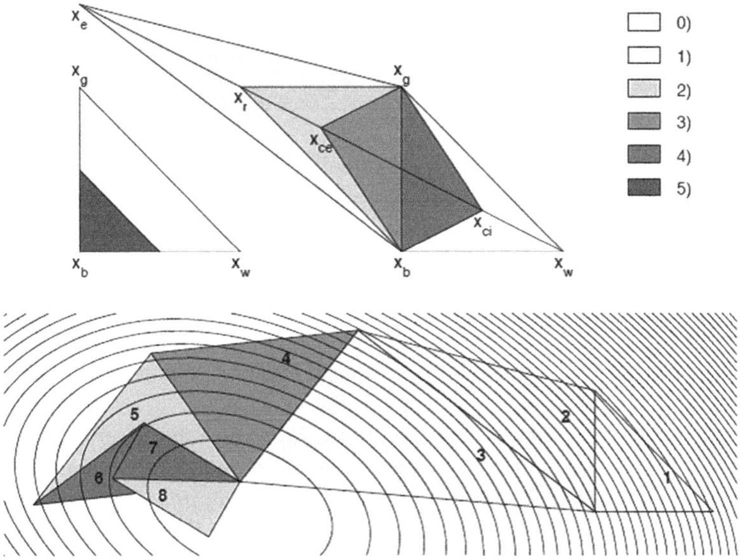

Fig. 4.1 Nelder-Mead method: (0) original simplex (triangle) (1) expansion, (2) reflexion (3) external contraction, (4) internal contraction, (5) reduction [5]

is the only one of vertices $\boldsymbol{x}_0, \boldsymbol{x}_1, ..., \boldsymbol{x}_n$, in which an objective function acquires the smallest value.

The first attempt to choose $\hat{\boldsymbol{x}}$, is called *reflection*, because a point

$$\boldsymbol{x}_r = \overline{\boldsymbol{x}} + (\overline{\boldsymbol{x}} - \boldsymbol{x}_w) \tag{4.1}$$

is the representation of a point $\boldsymbol{x}_w$ in central symmetry with the centre $\overline{\boldsymbol{x}}$, or flipping the top around the centre of gravity.

If $f(\boldsymbol{x}_r) < f(\boldsymbol{x}_b)$, then it means that a decrease of values on the half-line is considerable, and therefore we will try to move even further along this half-line $\boldsymbol{x}_w\overline{\boldsymbol{x}}$ to a point $\boldsymbol{x}_e$. The selection of a point $\boldsymbol{x}_e$ is called an *expansion* and we calculate it according to the relation:

$$\boldsymbol{x}_e = \overline{\boldsymbol{x}} + 2(\overline{\boldsymbol{x}} - \boldsymbol{x}_w) \tag{4.2}$$

If $f(\boldsymbol{x}_e) < f(\boldsymbol{x}_b)$, then $\hat{\boldsymbol{x}} = \boldsymbol{x}_e$, i.e., we replace the worst peak $\boldsymbol{x}_w$ with a point $\boldsymbol{x}_e$.

If $f(\boldsymbol{x}_e) \geq f(\boldsymbol{x}_b)$, so let us try to use at least a point $\boldsymbol{x}_r$. The condition for its inclusion in the simplex is the fulfilment of the condition $f(\boldsymbol{x}_r) < f(\boldsymbol{x}_g)$ for some vertex $\boldsymbol{x}_g$ other than the worst, i.e., for $\boldsymbol{x}_g \neq \boldsymbol{x}_w$ (index g resembles the English word *good*). If such a condition applies, we take $\hat{\boldsymbol{x}} = \boldsymbol{x}_r$ instead of the original point $\boldsymbol{x}_w$.

If neither $\boldsymbol{x}_e$ nor $\boldsymbol{x}_r$ satisfies, we try to find such a point $\hat{\boldsymbol{x}}$ on the line with the endpoints $\boldsymbol{x}_w, \boldsymbol{x}_r$, to $f(\hat{\boldsymbol{x}}) < f \min\{f(\boldsymbol{x}_w); f(\boldsymbol{x}_r)\}$. Specifically, we proceed as follows:

(a) if $f(\boldsymbol{x}_r) < f(\boldsymbol{x}_w)$, we try a point $\boldsymbol{x}_{ce} = \frac{1}{2}(\overline{\boldsymbol{x}} + \boldsymbol{x}_r)$ (it lies closer to a point $\boldsymbol{x}_r$), and
if $f(\boldsymbol{x}_{ce}) < f(\boldsymbol{x}_r)$, then $\hat{\boldsymbol{x}} = \boldsymbol{x}_{ce}$, so we will substitute $\boldsymbol{x}_w := \boldsymbol{x}_{ce}$;

(b) if $f(\boldsymbol{x}_r) \geq f(\boldsymbol{x}_w)$, we try a point $\boldsymbol{x}_{ci} = \frac{1}{2}(\overline{\boldsymbol{x}} + \boldsymbol{x}_w)$ (it lies closer to a point $\boldsymbol{x}_w$), and
if $f(\boldsymbol{x}_{ci}) < f(\boldsymbol{x}_r)$, then $\hat{\boldsymbol{x}} = \boldsymbol{x}_{ci}$, so we will substitute $\boldsymbol{x}_w := \boldsymbol{x}_{ci}$.

Selection of a point $\boldsymbol{x}_{ce}$ or $\boldsymbol{x}_{ci}$ is denoted as a *contraction* (in the lower index, the letter *c* resembles the word *contraction*; the letter *e* resembles the word *external*, i.e. $\boldsymbol{x}_{ce}$ lies outside the original simplex; and the letter *i* resembles the English word *internal*, i.e. $\boldsymbol{x}_{ci}$ it lies inside the original simplex).

If neither $\boldsymbol{x}_e, \boldsymbol{x}_r, \boldsymbol{x}_{ce}$ nor $\boldsymbol{x}_{ci}$ meet conditions, we assume that a vertex $\boldsymbol{x}_b$ in which an objective function acquires its smallest value, is close to a minimum.

Therefore, we perform a *simplex reduction*: a vertex $\boldsymbol{x}_b$ in a simplex remains and the remaining vertices $\boldsymbol{x}_i$ are shifted to the centre of the lines $\boldsymbol{x}_b\boldsymbol{x}_i$. Thus, simplex moves (retracts) towards the best vertex $\boldsymbol{x}_b$.

Transformation of a simplex, representing one step of the Nelder-Mead method, can be described shortly according to [5], as follows (Fig. 4.1):

(1) *expansion*:$f(\boldsymbol{x}_r) < f(\boldsymbol{x}_b)$ and $f(\boldsymbol{x}_e) < f(\boldsymbol{x}_b) \Rightarrow \boldsymbol{x}_w := \boldsymbol{x}_e$
(2) *reflexion*:$f(\boldsymbol{x}_r) < f(\boldsymbol{x}_g)$ for some vertex $\boldsymbol{x}_g \neq \boldsymbol{x}_w \Rightarrow \boldsymbol{x}_w := \boldsymbol{x}_r$.
(3) *external contraction*:$f(\boldsymbol{x}_r) < f(\boldsymbol{x}_w)$ and $f(\boldsymbol{x}_{ce}) < f(\boldsymbol{x}_r) \Rightarrow \boldsymbol{x}_w := \boldsymbol{x}_{ce}$.
(4) *internal contraction*: $f(\boldsymbol{x}_r) \geq f(\boldsymbol{x}_w)$ and $f(\boldsymbol{x}_{ci}) < f(\boldsymbol{x}_w) \Rightarrow \boldsymbol{x}_w := \boldsymbol{x}_{ci}$.
(5) *reduction*: $\boldsymbol{x}_i := \frac{1}{2}(\boldsymbol{x}_b + \boldsymbol{x}_i)$ for all $\boldsymbol{x}_i \neq \boldsymbol{x}_b$

We go gradually from the point (1) to the point (5) as shown in Fig. 4.1. If any of the conditions in the points (1)–(4) is not met, we move on to the next point. If it is fulfilled, then we implement a transformation, i.e., we replace a vertex $\boldsymbol{x}_w$ according to the given command. If the condition in none of points (1)–(4) is met, we perform a reduction according to the point (5). This method initially requires an initial approximation, i.e., the ***initial simplex*** $\boldsymbol{x}_0 = \left(\boldsymbol{x}_1^{(0)}, \boldsymbol{x}_2^{(0)}, \ldots, \boldsymbol{x}_n^{(0)}\right)^T$ and small number δ. The other vertices $\boldsymbol{x}_i$ of the initial simplex are derived from a vertex $\boldsymbol{x}_0$ by adding a number δ to its *i*-th component $\boldsymbol{x}_i^{(0)}$, i.e. $\boldsymbol{x}_i = \left(\boldsymbol{x}_1^{(0)}, \ldots, \boldsymbol{x}_i^{(0)} + \delta, \ldots, \boldsymbol{x}_n^{(0)}\right)^T$, $i = 1, 2, \ldots, n$.

We repeatedly transform the simplex. We end the calculation and consider a vertex $\boldsymbol{x}_b$ to be a sufficiently good approximation of a minimum $\boldsymbol{x}^*$, if the vertices of the simplex are close enough to each other and functional values in them differ very little, i.e., when with the specified tolerance $\varepsilon_1, \varepsilon_2$ the following is valid

$$\|\boldsymbol{x}_i - \boldsymbol{x}_b\| < \varepsilon_1 \;\wedge\; |f(\boldsymbol{x}_i) - f(\boldsymbol{x}_b)| < \varepsilon_2 \, for \, \forall \text{ vertice } \boldsymbol{x}_g \neq \boldsymbol{x}_w \tag{4.3}$$

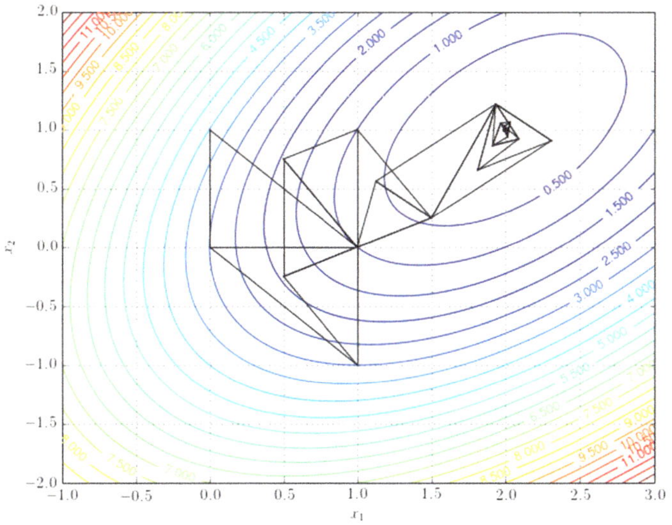

Fig. 4.2 Nelder-Mead method, sequence of iterative steps [11]

The Nelder-Mead method is a *heuristic method* (a heuristic or discovery method is a procedure that is based not only on logical reasoning and experience, but also on observation and experimentation).

This method is suitable for minimizing the functions of a smaller number of variables, i.e., $n \leq 10$. Although very little is known about the convergence of this method, practice speaks in its favour where the method is surprisingly very successful. Therefore, it is considered a reliable or robust method. A major disadvantage of the method is that it is slow, especially when close to the minimum. Another minus is the large volume of calculations, but it is still not a "dead method", which is obvious also from the fact that it is implemented as a basic method of multi-dimensional minimization in MATLAB core function `fminsearch`. Examples of the Nelder-Mead method for finding an extremum of a selected function are shown in Figs. 4.2 and 4.3.

4.1.2 Cauchy Steepest Descent Method

The *Steepest Descent* method is one of the basic minimization methods using a derivation of an objective function. The simplest numerical method of finding a minimum of a constraint function is based on the idea of finding an optimum in a descent direction of a function. If a function $f(x)$, $x \in \mathbb{R}^n$ is differentiable, then its negative gradient $-\nabla f(x^k)$ indicates the *direction of the steepest descent* of a function f at a point x^k. In Definition 3.1.1, we have defined the *descent direction* by the relation (3.23). If we choose descent direction as the negative gradient in the model algorithm (Table 4.1), thus the so-called *Cauchy direction* (3.25) $s^k = -\nabla f(x^k) = -g_k$ and a

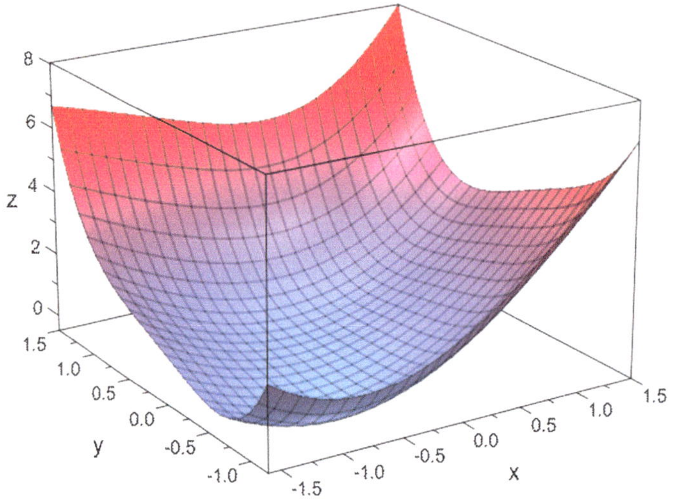

Fig. 4.3 For a minimum of a function $f(x, y) = x^2 + y^4 + x - \sin(xy)$ in $\mathbb{R}^3$ is valid $f(\boldsymbol{x}^*) = f(-0.754; -0.5563) = -0.497$

step λ_k is selected to be optimal, i.e., as a solution of the one-dimensional problem (3.22), then we obtain the classical *Cauchy steepest descent method* [5, 12, 13].

An optimal step size λ_k for the k-th iteration occurs at a minimum value of a function $\varphi(\lambda)$, which is the solution to the problem (3.22): $Min\{\varphi(\lambda) := f(x^k + \lambda s^k) \mid \lambda \in \mathbb{R}\}$. Instead of calculating the exact value of a minimum λ_k we only calculate an approximation using few steps of a suitable one-dimensional minimization method, e.g., using Davidon's cubic interpolation. Exact minimization would be an unnecessary luxury, as we only determine the intermediate result x^{k+1} on the way to a minimum x^* using λ_k. Simply put, at the beginning of each iteration step we are at a point x^k and want to get closer to a minimum. So, we choose a direction s^k (descent direction) in which the value of a function f klesá, decreases, and on the semi-line $x^k + \lambda s^k, \lambda > 0$ we select a point

$$x^{k+1} = x^k + \lambda_k s^k, \tag{4.4}$$

Table 4.1 Model solution algorithm

Input	Objective function $f(x)$ of the problem (U1) Initial point of iteration process x^0 Tolerance constant $\varepsilon > 0$ for accuracy criterion Iteration counter settings: $k = 0$
Calculation	(1) Accuracy testing. If $\left\|\nabla f(x^k)\right\| < \varepsilon$, finished (2) Calculation of descent direction s^k (3) Step length calculation λ_k (4) New approximation $x^{k+1} = x^{k+1} + \lambda_k s^k$ (5) Cycle repeat. $k = k + 1$ go to (1)

in which $f\left(x^{k+1}\right) < f\left(x^k\right)$ will apply. The number λ_k is usually called a *step length parameter* if a direction vector s^k is unitary, i.e., if $\left\|s^k\right\|_2 = 1$, then $\lambda_k = \left\|x^{k+1} - x^k\right\|_2$ is the distance of points x^k and x^{k+1}, i.e., if λ_k is a step length. If we did not change a step length, we would have a gradient method with a constant step. We require an optimal step λ_k, which we can also express as follows:

$$\lambda_k \approx \min_{\lambda>0} \varphi(\lambda), \quad \text{where} \quad \varphi(\lambda) = f\left(x^k + \lambda s^k\right) \tag{4.5}$$

We want such λ, that meets the condition

$$\varphi(\lambda) \leq \varphi(0) + \mu\, \varphi'(0), \tag{4.6}$$

where $\mu \in (0, 1)$ is the selected constant. The condition (4.6) is called the *Armijo-Goldstein condition.* It turned out that estimation satisfying this condition is sufficient for iteration in the algorithm, in contrast to the exact solution, which in many cases is more difficult to calculate.

There are several ways to choose a criterion for terminating iterations. Since a gradient of a function f must be zero in the optimum, a suitable criterion for terminating the calculation is $\left\|\nabla f\left(x^{k+1}\right)\right\| < \varepsilon$ for the appropriately selected $\varepsilon > 0$, what is the prescribed tolerance, as seen in Table 4.1. If this condition is met, a point x^{k+1} can be considered as a satisfactory approximation of a minimum x^*. Similarly, a suitable criterion for terminating calculations is if iterated solutions become close enough, i.e., $\left\|x^{k+1} - x^k\right\| \leq \varepsilon$, for some $\varepsilon > 0$.

Figure 4.4 graphically represents the principle of descent methods. In initial stages of calculations, when we are still quite far from a minimum, there is usually a relatively rapid decrease in the values of an objective function. On the other hand, near a minimum, convergence is already slow, only linear, i.e., $\left\|x^{k+1} - x^*\right\| \leq C\left\|x^k - x^*\right\|$, where the constant is less than one, but often only slightly.

Fig. 4.4 Principle of descent methods [5]

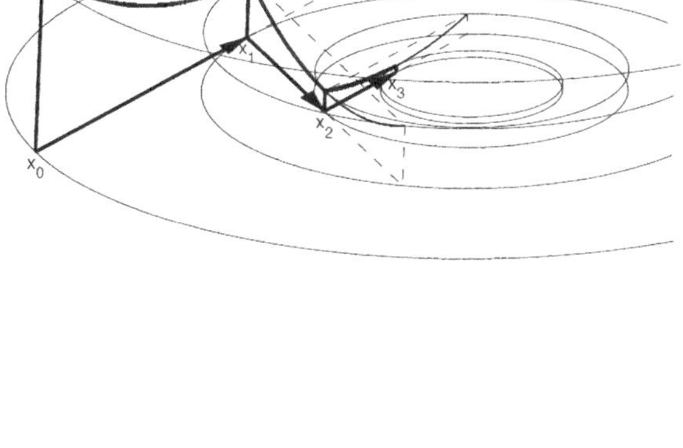

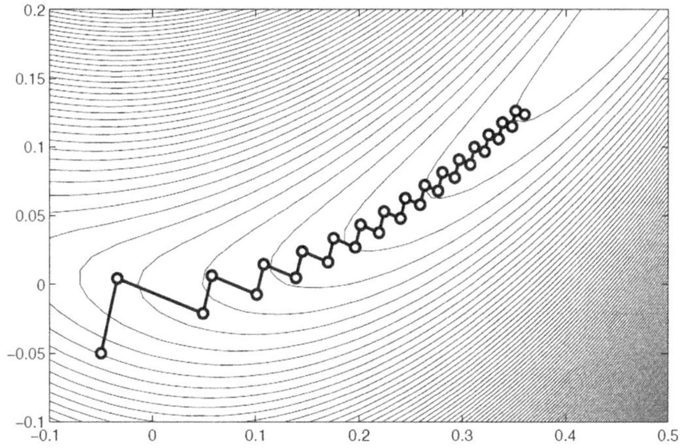

Fig. 4.5 Zig-zag effect in steepest descent method [5]

Directions of the steepest descent in iterations change the sign because the direction of the steepest descent is always perpendicular to the contours of a function (it follows from the gradient property). Since the direction vectors of the individual iterations x^{k+1} and x^k are perpendicular to each other, the iterated solutions, or directions of movement to the optimum are therefore quite winding, thus it is called the zig-zagging effect as presented in Fig. 4.5. Near the minimum, the trajectory of movement towards the minimum is ridiculous, it is formed by increasingly shorter sections perpendicular to each other. It is this feature that explains the slowness of this method.

4.1.3 Newton's Method

It is a gradient method that tries to find a minimum x^* as a solution of a system of nonlinear equations in a matrix form $g(x) = 0,\ x \in \mathbb{R}^n$ using Newton's method. Within older vector notation (bold font), the system has a matrix shape $g(\boldsymbol{x}) = 0,\ \boldsymbol{x} \in \mathbb{R}^n$. Such a solution can be any stationary point of a function f, i.e. a point $x^* \in \mathbb{R}^n$, at which the value of a function gradient f is zero, i.e., $\nabla f(x^*) = 0$. Obviously, we can also find a maximum or a saddle point.

It is a second-order method, so let us consider that an objective function of $n-$variables $f(x)$ has continuous second partial derivatives and can be approximated by the second degree Taylor polynomial according to the relation (3.5) with neglect of the remainder. The remainder can be written as follows: $f(\boldsymbol{x} + \boldsymbol{p}) = f(\boldsymbol{x}) + g(\boldsymbol{x})^T \boldsymbol{p} + \frac{1}{2} \boldsymbol{p}^T \boldsymbol{H}(\boldsymbol{x}) \boldsymbol{p}$, where $\boldsymbol{p}$ expresses a vector change, $g(\boldsymbol{x})$ is a gradient and $\boldsymbol{H}(\boldsymbol{x})$ is a Hessian of a function f. Let us use the current notation, i.e., the gradient of a function $f(x),\ x \in \mathbb{R}^n$ will be denoted by the symbol $g(x)$, the Hessian matrix of the second partial derivatives by the symbol $H(x)$, the change vector simply p. Thus,

$g(x) := \nabla f(x)$, $H(x) := \nabla^2 f(x)$. If we assume that $x^k \in \mathbb{R}^n$ it is an approximate solution of the problem (U1) and we express the vector change in the form $p = x - x^k$, then we approximate a function $f(x)$ around a point x^k by the second degree Taylor polynomial in the following form:

$$f(x) \cong f_k + g_k^T\left(x - x^k\right) + \tfrac{1}{2}\left(x - x^k\right)^T H_k\left(x - x^k\right) =: Q(x), \tag{4.7}$$

where $f_k = f\left(x^k\right)$, $g_k = f\left(x^k\right) = \nabla f\left(x^k\right)$, $H_k = H\left(x^k\right) = \nabla^2 f\left(x^k\right)$,
or simply

$$f(x) \cong f_k + g_k^T p + \frac{1}{2} p^T H_k p \tag{4.8}$$

In Newton's method, a new approximate solution $x^{k+1} \in \mathbb{R}^n$ of the problem (U1) is defined as a minimum of a quadratic function $Q(x)$, $x \in \mathbb{R}^n$ defined by the relation (4.7). Assuming a positive definite matrix H_k, a function $Q(x)$ will be strongly convex. It follows that it will have the only minimum that we find by fulfilling the necessary condition for the existence of a local extremum, i.e., a function gradient $Q(x)$ must be zero [13]. By derivation we get the following system of equations:

$$\nabla Q(x) = g_k + H_k\left(x - x^k\right) = 0, \tag{4.9}$$

after adjustment we get $x - x^k = -H_k^{-1} g_k$, or

$$x^{k+1} = x^k - H_k^{-1} g_k \tag{4.10}$$

The algorithm of Newton's method is derived from the iteration scheme, see Table 4.2.

Newton's method is a special case of the model algorithm given in Table 4.1 in which we define the descent direction by the following relation

Table 4.2 Algorithm of Newton's method

Input	Objective function $f(x)$ of the problem (U1) Initial point of the iteration process $x^0 \in \mathbb{R}^n$ Tolerance constant $\varepsilon > 0$ for the accuracy criterion Iteration counter settings: $k = 0$
Calculation	(1) Gradient calculation $\nabla f(x^k) = g_k$, if $\|g_k\| < \varepsilon$, finished (2) Hessian matrix calculation $\nabla^2 f(x^k) = H_k$ and solving a system of equations $H_k p = -g_k$, its solution we will denote p_k (3) Calculation of a new iteration $x^{k+1} = x^k + p_k$ (4) $k = k + 1$, go to (1)

$$s^k = p_k = -H_k^{-1} g_k, \tag{4.10*}$$

which is *Newton direction,* and we leave the step length constant in each iteration, i.e., $\lambda_k = 1$. According to the Consequence 3.1.2, Newton direction (vector) $p_k \in \mathbb{R}^n$ is a descent direction in the sense of the Definition 3.1.1 only under the assumption of a positive definiteness of the Hessian matrix H_k [1, 13].

The advantage of Newton's method is a quadratic convergence rate and $\lambda_k = 1$ in each iteration step, which means that we are not burdened with solving the problem of optimal step length. Disadvantages of Newton's method include the dependence of convergence properties on the selection of an initial point, the need to solve Hessians H_k in each iteration, and then the need to solve systems of equations. In addition, the constancy of a step does not guarantee the descentness of the method in the sense of the relation (3.15) and therefore it is recommended to regulate the step to the optimal length. Efforts to eliminate the main disadvantages of Newton's method gave impetus to the development of the so-called *modified Newton's method*, which uses *Newton direction* p_k given by the relation (4.10*), but uses the optimal step length, or approximately optimal. The algorithm of *modified Newton's method* can be expressed in Table 4.2, the difference from the algorithm of the ordinary Newton's method will be only in the point (3), which is modified to: the calculation of the step length λ_k and calculation of a new iteration $x^{k+1} = x^k + \lambda_k p_k$. If Newton's method converges, its rate is very good. Although Newton's method has excellent convergence properties near the solution, it is sometimes not appropriate to use it. Especially when it comes to problems with many unknowns, when it is very difficult to calculate the Hessian in each iteration. In such cases, instead of accurately calculating the Hessian matrix, only the Hessian approximation is preferred. Methods with this approach are known as *quasi-Newton methods.*

4.1.4 Quasi-Newton Methods—BFGS

The impetus for the emergence of these methods was the effort to eliminate the main disadvantages of Newton's method (an effort to eliminate the need to compute the Hessian and solve a system of linear equations in each iteration). These methods are significant modifications of Newton's method and belong to the category of the most used methods in unconditional optimization. In addition, they are ambitious in their efforts to minimize the dependence of convergence on the selection of the initial point. These are two-step gradient methods that implement an approximate Newton's iteration and at the same time specify a matrix that appears in the iteration as an approximation of the (inverse) Hessian matrix. The basic idea of quasi-Newton methods lies in gradual approximation of the Hessian matrix or inverse Hessian matrix using gradients $g_k = \nabla f(x_k)$. If the inversion of the Hessian matrix and not the Hessian matrix itself is approximated, there is no need to solve systems of linear equations $H_k p = -g_k$, which means that both above-mentioned disadvantages of Newton's method are eliminated in this way. Individual quasi-Newtonian methods differ only in the so-called *quasi-newton formulas*, which in an iterative step describe the transition from one approximation of the (inverse) Hessian matrix at a point $x^k \in \mathbb{R}^n$ of a function f s gr with a gradient $g_k = \nabla f\left(x^k\right)$ to the improved approximation of the (inverse) Hessian matrix at the new point $x^{k+1} \in \mathbb{R}^n$ based on new information $g_{k+1} = \nabla f\left(x^{k+1}\right)$.

Quasi-Newton formulas (QNF), in which matrices B_k approximating the Hessian matrix appear, are called *direct QNF*.

Quasi-Newton formulas (QNF), in which matrices G_k approximating the *inverse* Hessian matrix appear, are called *indirect QNF*.

The improved approximation B_{k+1} can be expressed through additive matrix correction B_k, i.e., in the form $B_{k+1} = B_k + \Delta B_k$, or $B_{k+1} = B_k + U_k$. The matrix ΔB_k and the matrix U_k, both with low rank of matrix, is called the *correction matrix* or the *update matrix*, respectively. Of course, these matrices must have certain specific properties resulting from the very algorithmic nature of quasi-Newton methods (mimicking Newton's iteration) and from the implementation aspects of the algorithm (simplicity requirement). Thus, the following is required: 1. symmetry of matrices B_k; 2. positive definiteness of matrices B_k; 3. the simplicity of the correction matrix ΔB_k expressed by its low rank (usually with a rank of one or two). The first requirement is obvious, since the Hessian matrix is symmetric, its inverse matrix will also be symmetric. The second requirement follows from the descent direction requirement (Lemma 3.1.3). The third requirement creates a presumption of expressing the correction matrix ΔB_k by a simple formula. The same can be written for improved approximations G_{k+1} of the inverse Hessian matrix.

In quasi-Newton methods, a function f is approximated by

$$f(x+p) \simeq \hat{f}(p) = f(x) + g(x)^T p + \tfrac{1}{2} p^T B(x) p, \ \ x \in \mathbb{R}^n, \tag{4.11}$$

where a matrix $B(x)$, hereinafter referred to only as B, represents an approximation of the Hessian H. Using the Definition 2.1.5 it can be found that a function $\hat{f}(p)$ is strictly convex and therefore has one minimum, which, similarly to Newton's method, can be found from the fulfilment of the necessary condition $\nabla \hat{f}(p) = 0$, which leads us to the following form

$$p = -B^{-1} g, \tag{4.11*}$$

representing the descent direction. This is a Newton *descent direction*, which in $k-$iteration will be denoted as $s_k = -B_k^{-1} g_k$. After calculating the decsent direction, a new point $x^{k+1} = x^k + \lambda_k s_k$ is calculated where an optimal step length can be calculated in many ways. The Armijo reverse search algorithm is often applied, or calculations are performed using stronger Wolfe conditions [1, 13]. For the new point, a gradient g_{k+1} and vectors $y_k = g_{k+1} - g_k$, $p_k = x^{k+1} - x^k$ are calculated by means of which we calculate the correction matrix ΔB_k and finally the new matrix B_{k+1} in k-th iteration step. By modification, the relation can be obtained for the new matrix

$$B_{k+1} p_k = y_k, \tag{4.12}$$

which is a system of equations calling a *quasi-Newton condition.* At iteration points x^{k+1} and x^k, the similarity of gradients is required for a function f and its approximation $\hat{f}$, if in addition we require a positive definiteness of the matrix B_{k+1} at these points, the *curvature condition* $p_k^T y_k > 0$ will apply. The following theorem is related to the curvature condition, which is known as the first theorem on curvature validity.

Theorem 4.1.1 *If calculation of the step length* λ_k *is accurate and if the matrix* B_k *is positive definite, then the curvature condition applies.*

The quasi-Newton condition can be satisfied in several ways, most often by using the following relation

$$B_{k+1} = B_k + \nabla B_k(p_k, y_k, B_k) \tag{4.13}$$

which is the mentioned additive correction.

Calculation of the improved Hessian approximation, i.e., calculation of the matrix B_{k+1} vedie k riešeniu sústavy lineárnych rovníc leads to the solution of a system of linear equations. We want and can avoid this by calculating approximations of the inverse Hessian matrix, i.e., the matrix G_{k+1}, then $G = B^{-1}$ is valid when denotation $B_k \approx [\nabla^2 f(x^k)]$ and $G_k \approx [\nabla^2 f(x^k)]^{-1}$ is used. The validity of the theorem is known, stating that if a square and regular matrix is positive definite, then its inverse matrix is also positive definite. Based on this, we can implement approximation G_{k+1} without worrying about the loss of positive definiteness of the matrix. The above facts can be clearly summarized into a general algorithm of quasi-Newton methods, which is expressed in Table 4.3, where the calculation of one, specifically $(k+1)$-th iteration of quasi-Newton method (QNM) is presented.

Table 4.3 Algorithm of **Quasi-Newton method (QNM),** $(k+1)$-th **iteration**

Input	Objective function $f(x)$ of the problem (U1) Initial point of the iterarion process $x^k \in \mathbb{R}^n$ Gradient $g_k = \nabla f(x^k)$ Symmetrical, positive definite matrix $G_k \approx \left[\nabla^2 f(x_k)\right]^{-1}$
Calculation	(1) We calculate the decsent direction $s_k = -G_k g_k$ (2) In direction s_k we calculate the optimal step λ_k (3) We calculate the new point $x^{k+1} = x^k + \lambda_k s_k$ (4) We calculate gradient g_{k+1} and vectors $y_k = g_{k+1} - g_k, p_k = x^{k+1} - x^k$ (5) Using vectors p_k, y_k and matrix G_k We construct the correction matrix ΔG_k satisfying the quasi-Newtonian condition (6) We calculate the new matrix $G_{k+1} = G_k + \Delta G_k$

Historically, the first quasi-Newton method was the method proposed by Davidon in 1959 [14], developed in 1963 by Fletcher and Powell. Therefore, it is known as DFP.

Quasi-Newton BFGS formula was derived in different ways in 1970–71 by four mathematicians: Broyden, Fletcher, Goldfarb, and Shanno, after whom it was named. Many years of numerical experiments have shown that the BFGS method generally converges better than the DFP method. Using the convention that we write the quantities of $k-$th iteration without an index and the values of the quantities $(k+1)-$th iteration we write with a subscrip+, so the so-called direct BFGS formula can be written in the following form:

$$\Delta B_{BFGS} = \frac{yy^T}{y^T p} - \frac{Bpp^T B}{p^T Bp}, \tag{4.14}$$

and complement to the inverted BFGS formula in the form as follows

$$\Delta G_{BFGS} = \left(1 + \frac{y^T Gy}{p^T y}\right)\frac{pp^T}{p^T y} - \frac{Hyp^T + py^T G}{p^T y} \tag{4.15}$$

which is the direct DFP formula.

Calculation of the approximated matrix G_{k+1} can be expressed in the form $G_{k+1} = G_+^{BFGS} = G + \Delta G_{BFGS}$, or in the break down form

$$G_{k+1} = \left(I - \frac{p_k y_k^T Gy}{p_k^T y_k}\right) G_k \left(I - \frac{y_k p_k^T}{y_k^T p_k}\right) + \frac{p_k p_k^T}{p_k^T y_k} \tag{4.16}$$

where I denotes a unit matrix. We see that in the formula (4.16) each value of the participating quantities contains index k and therefore this index has no informative value and for this reason the convention of omitting the index is mentioned. To start the quasi-Newton method, in addition to the initial point x^0 (and the corresponding gradient g_0 in it), we must also enter the initial matrix G_0. As a rule, $G_0 = I$ is selected, i.e., the first iteration will coincide with the Cauchy iteration.

It is desirable that the sequence maintain a positive definiteness. The condition for the transfer of positive definiteness by individual iterations for the BFGS method is guaranteed by the following theorem.

Theorem 4.1.2 *Assuming the validity of the curvature condition, the BFGS formula retains a positive definiteness.*

Finally, just a note on the selection of the termination criterion. It is usually selected in the form $\|g\| < \varepsilon$ which can be understood that a given algorithm operates if the standard from the gradient g is not less than the selected tolerance ε.

4.2 Selected Methods of Linear Programming

The impetus for the birth and development of linear programming was the effort to optimize military processes during World War II. American mathematician G. B. Dantzig is a father of linear programming. In 1947 he formulated and published the linear programming problem and the method of its solution. However, it turned out that Russian mathematician Kantorowich, who solved optimization problems in the plywood industry and later optimized military processes during the Leningrad blockade. He solved the optimal distribution of military equipment due to the thickness of ice during the supply of Leningrad across Lake Ladoga in 1933. Regardless of the primacy, the birth of linear programming as a mathematical model and its further development was important (J. Neumann, W. Leontief, T. Koopmans). Linear programming was generally at the beginning of mathematical programming and over time the problems of quadratic, convex, stochastic, nonlinear, and other types of mathematical programming were formulated, see [15, 16].

While the development of mathematical programming problems took place in the line from simpler to more complex, from concrete to abstract, with current knowledge in the field we know that linear programming is only a special case of nonlinear and convex programming, i.e., is a subset of them as expressed by the scheme $LINEAR \subseteq CONVEX \subseteq NONLINEAR$. Therefore, we first introduced the methods of nonlinear programming. Now let us take a closer look at the methods of linear programming.

4.2.1 Simplex Method

The simplex algorithm, or the simplex method, is the first algorithm introduced by G. B. Dantzig in 1947 to solve the problem of linear programming. This method was even ranked by *The Journal Computing in Science and Engineering* in the list of the 10 best algorithms of the 20th century. Detailed theory and the detailed simplex algorithm can be found in [1, 3, 16]. Our goal is to briefly recall the basic principle of the simplex method.

Let us consider a linear programming problem in standard form, i.e., the SLP presented in Sect. 2.3.1 as $f(x) = c^T x \to \min$, $Ax = b$, $x \geq 0$. Let us suppose we can write a matrix $A \in \mathbb{R}^{m \times n}$ in the form $A = [B, N]$, where $B \in \mathbb{R}^{m \times m}$ is a matrix with rank m. If we use a similar notation for the vectors $x = [x_B, x_N]$ and $c = [c_B, c_N]$, hen after short modifications we get an equivalent notation for the constraints:

$$x_B + B^{-1} N x_N = B^{-1} b \tag{4.17}$$

A solution that satisfies $x_N = 0$ and $x_B = B^{-1}b$ is called a *basic solution.* If $x_B \geq 0$ we call it a *basic feasible solution.* Let us use $Z = c^T x$ for an objective function. Then the following notations are equivalent:

$$\begin{aligned} Z &= c^T x \\ Z - \left(c_N - c_B B^{-1} N\right) x_N &= c_B B^{-1} b \end{aligned} \tag{4.18}$$

In the theory of linear programming, an important finding has been reached. If a LP problem has an optimal solution, it must have an optimal basic feasible solution. To find an optimal solution, it is enough to examine basic feasible solutions. There are many such solutions presented in [9]. The simplex algorithm gives us instructions on how to go through basic feasible solutions so that the value of an objective function is gradually reduced. The significance of the last equality in the (4.18) lies in the fact that if $c_N - c_B B^{-1} N \leq 0$, then the basic solution $x_B = B^{-1}b, x_N = 0$ is the optimal solution.

The simplex algorithm always finds an optimal solution (if any exists), but the important issue is what demands this algorithm places on calculations. Linear programming is a special case of convex mathematical programming, but since its inception in 1947 it developed completely independently and in isolation from other branches of mathematical programming. It has not been known for a long time whether a simplex algorithm belongs to the class of the so-called *non-polynomial algorithms*, or *polynomial* ones that are able to find an optimal solution in polynomial time. The simplex method was used to be consider as universal, reliable, and powerful method, enabling routine solving of linear programming problems, therefore any other attempts to solve LP problems by non-simplex approaches were viewed with distrust. Certain nonlinear procedures for solving LPs have even gone unnoticed, e.g., Frisha's work in 1955 [17]. The change took place in the 60's and 70's, when a new branch was created dealing with computational complexity of algorithms. Scientists preferred the idea that every fast program should be *time polynomial*, which means that the number of arithmetic operations needed to solve a dimensional problem m can be estimated from above by a polynomial in a variable *m*. In 1972, V. Klee and G. Minty created the example of the LP problem and showed that the simplex method is not polynomial, because in the worst cases it takes an exponentially many iterations to find an optimal solution. This has caused concern in the professional community and at the same time a better will to take serious alternative approaches to solving LP problems. In 1979, Russian mathematician Leonid Khachiayan proved that an *ellipsoid algorithm* (a type of subgradient method using some nonlinear optimization methods) solves an LP in less than polynomial time depending on the dimensions of the matrix *A* and the number of digits on the input. Nevertheless, this algorithm has proven inefficient for engineering practice problems. Later Khachian's algorithm has proved hopelessly slow for common LP problems. Therefore, the Simplex method reigned in linear programming for another few years. In the mid-1980s, however, there was a dramatic change [17–19].

As mentioned above, in 1984 Indian mathematician Narendra Karmakar introduced the so-called projective algorithm with the later used name of an *interior point* algorithm and proved that it solves a LP problem in polynomial time. This algorithm, in contrast to the ellipsoid one, proved to be very effective for practical tasks with a high number of variables. So, the next chapter will be devoted to some of the interior point methods. We will present the form of this algorithm for linear programming. We will also deal with this algorithm in connection with quadratic programming. See more in [1, 15, 18–21].

4.2.2 *Interior Point Methods for Linear Programming*

Interior point methods (IPM) were widely cultivated and analysed especially in the 1960s within nonlinear programming (NP) but declined over time due to failure in applications to solve NP problems. Interest in these methods was renewed in 1984 with the publication of the so-called Karmarkar's project algorithm. In 1986, experts [21] have found that this algorithm was very closely related to the *logarithmic interior point barrier method*, which was also unsuccessfully applied to solving NP problems in the 1960s. They again drew the attention of many mathematicians to IPM and mathematical methods. Programming began to develop very rapidly. The *era of modern interior point methods in optimization* has begun. Several types of internal point methods for linear programming have been developed. Unlike the simplex method, iterations take place inside the feasible region and not along the endpoints. IPM is a very good tool for LP problems, but it has been shown to be able to solve other types of problems such as quadratic programming (QP) problems. It solves convex QP problems even in polynomial time. It can be said that IPM is a modification of Newton's method with the fact that it can work with constraints [17].

The basic idea of interior point methods (IPM) is based on solving the problem of convex programming (U3),

$$Min\{f_0(x) \mid x \in X,\ f_i(x) \leq 0,\ i = 1, 2, ..., m\} \quad \text{(U3)}$$

i.e., the problem (U3) mentioned in Sect. 2.3.4, where all functions f_i, $i = 0, 1, ...m$ are convex. Let consider the problem (U3) with an optimal solution $\hat{x}$ and a feasible region $K_3 = \{x \in \mathbb{R}^n \mid f_i(x) \leq 0, i = 1, 2, ..., m\}$, which we assume to contain some interior point, i.e., it meets the so-called *Slater condition*:

$$K_3^0\left\{x \in \mathbb{R}^n | f_i(x) < 0,\ i = 1, 2, \ldots, m\right\} \neq \emptyset \tag{4.19}$$

The points from K_3^0 are called *interior points* and the set K_3^0 is called the *interior of feasible region*. In this case, K_3^0 is the relative interior of K_3 and the closure of K_3^0 is K_3. The border is formed by those feasible regions $x \in K_3$ for which there exists such i so that $f_i(x) = 0$.

The fact that the optimal solution may be at the border of the feasible region does not allow the direct use of standard free optimization techniques. The starting point of the situation is the construction of the so-called *transformation barrier functions*, which became the basis of IPM. What is its essence?

The whole *class of auxiliary parameterized problems* (U3r), parameterized by a parameter $r > 0$, is assigned to the problem (U3):

$$Min_x\{T(x, r) \mid x \in K_3^0\} \quad (\text{U3}r)$$

In the parameterized problem (U3r), the new *objective function* $T(x, r) : K_3^0 \times \mathbb{R}_{++} \rightarrow \mathbb{R}$ appears. It is called the *transformation barrier function*, which for the practical purposes usually has the following form:

$$T(x, r) = f_0(x) + r\sum_{i=1}^{m} B(f_i(x)) \tag{4.20}$$

It is constructed using the so-called barrier *function* $B(y) : \mathbb{R}_{--} \rightarrow \mathbb{R}$ with the following properties (b1)–(b3):

$$\begin{aligned} &(\text{b1}) \lim_{y \to 0_-} B(y) = +\infty; \quad \text{b2}) \, \forall y < 0 : B'(y) > 0; \\ &(\text{b3}) \, \forall y < 0 : B''(y) > 0 \end{aligned} \tag{4.21}$$

The property (b1) represents a *barrier property* that causes the values $B(y)$ to increase indefinitely for y approaching zero, and at the same time this property implies that the function $T(x, r)$ is able to create a barrier at the border of the set K_3^0. The other two properties (b2), (b3) ensure that the objective function $T(x, r)$ in the auxiliary problem (U3r) is a convex function, which implies its continuity and the existence of a minimum in the set K_3. The following applies to the sequence of a free minimum $x(r)$ of the functions $T(x, r)$, thus the following is valid:

$$\lim_{r \to 0^+} x(r) = \hat{x}, \tag{4.22}$$

which implies that for a sufficiently small parameter $r > 0$, the free minimum of the objective function $T(x, r)$ sufficiently approximates the optimal solution $\hat{x}$ of the original problem (U3). In other words, the transformed objective function $T(x, r)$ has as its domain the interior of the feasible regions K_3^0 and its values increase indefinitely as we approach the border from the interior of the feasible region. Thus, if this function has a finite infimum, this infimum is a minimum that lies within a feasible region and free optimization methods can already be used to find it [22].

In practice, the algorithm based on gradual solution of parameterized problems (U3r) by some free optimization method from a given initial point is preferred, where we choose the initial value of a parameter $r > 0$ appropriately. Subsequently, the value of a parameter r is reduced via the reduction factor while the optimal solution from the previous problem is determined as the initial point for the problem with the new (reduced) parameter r.

The algorithm of the method can be schematically expressed in Table 4.4. A clear illustration of the IPM method is offered by R. J. Vanderbei in the publication [18] on the example of *a linear two-dimensional problem*. See Fig. 4.6.

Table 4.4 Interior Point Methods (IPM) algorithm

Input	Functions f_i, $i = 0, 1, \ldots, m$ defined on $\mathbb{R}^n$ Initial point $x^0 \in \mathbb{R}^n$ $r_1 > 0$ initial value of the parameter $\alpha \in (0, 1)$—reduction factor
Algorithm	$k = 1$ If the stop criterion is not met – Following from the point x^{k-1} we will find theminimum of the function $T(x, r_k)$ on K_3^0 – If x^k is a good approximation of $\hat{x}$ then finished – otherwise $r_{k+1} = \alpha\, r_k$

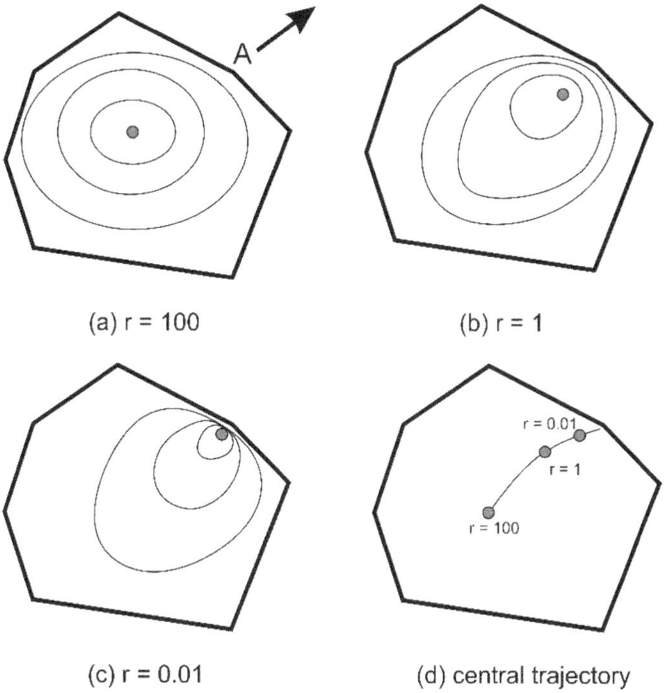

Fig. 4.6 **a–c** Contour line of the logarithmic transformation function for three different parameter values r in the LP problem; **d** central trajectory—curve of optimal solutions $x\,(r)$

Solutions $x(r)$, i.e., the *minima of the parameterized problems* (U3r) for decreasing parameter values r create a curve that passes inside the feasible region and ends in the optimal solution of the original problem. This curve is called the *central path* and is the central concept of interior point methods. It can be expressed as a set of points

$$C = \left\{x(r) \in K_3^0 \mid r > 0\right\} \quad (4.23)$$

A detailed examination of the *properties of the central trajectory* was important for proving the convergence properties of the generated IPM algorithms. To solve auxiliary barrier problems (U3r), the so-called Newton's method with shortened step length is used. Figure 4.6 illustrates how the minima $x(r)$ of transformation problems (U3r) for three different parameter values r form a central trajectory.

In the Fig. 4.7 we can see the calculated approximate *minima of transformation functions* $x(r)$ (white circles) which can be obtained by the free optimization method after a certain number of iterative steps. The vertices of the dashed polyline show the iteration points of the solution of the auxiliary parameterized problems. Especially within modern methods of the interior point, the modified Newton's method that is also called Newton's method with shortened step length, is the most frequently used free optimization method.

If we compare *old and new approaches* to interior point methods, we can perceive *classical & modern IPM* while the differences are manifested in their four characteristics.

- *Problem structuring:* old approaches applied IPM methods to very general and unstructured NP problems which made it difficult to detect certain regularities. It is characteristic of new approaches that they were first applied and analysed on simple LP problems of the following type

$$(LP)\ \ Min_x\left\{c^T x \mid Ax = b,\ \ x \geq 0\right\}, \quad (4.24)$$

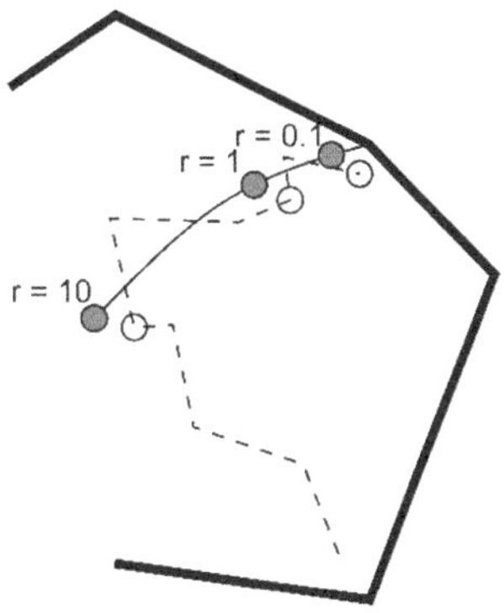

Fig. 4.7 Black circles—exact solutions $x\,(r)$ of auxiliary parameterized problems for three different parameter values $r = 10,\ r = 1,\ r = 0.1$. White circles—calculated approximate solutions

where $c, x \in \mathbb{R}^n$, $b \in \mathbb{R}^m$ and is a given matrix. The linearity of the problems as well as simple and symmetrical dual relations of linear programming were used and basic regularities in the LP problems were revealed. For this purpose, IPM have also been applied to some nonlinear convex problems with a certain structure and with well-described dual relationships.

- *Logarithmic transformation function.* In classical approaches, experiments were performed with the selection of a barrier function to achieve the best possible behaviour of algorithms, e.g., the Fiacco-McCormick barrier $B(y) = (-y)^{-1}$. In modern IPM, the *logarithmic barrier function* $B(y) = -\ln(-y)$ is preferred for its good property that can be applied to both primary and dual problem of LP while maintaining duality.
- *Newton's method.* While in old approaches various free optimization methods were used in IPM, in modern IPM the modified Newton's method is used, i.e., Newton's method with shortened step length. It was possible to determine the surroundings of the central trajectory with guaranteed quadratic convergence of Newton's method which enabled *free tracking* of the central trajectory in algorithms and acceleration of calculations.
- *Free tracking of the central trajectory.* The fourth characteristic feature of distinguishing modern IPM. In old approaches, the convergence of only the exact minima of transformed problems was proved and the effort to follow the central trajectory as accurately as possible was one of the sources of numerical instability in the implementation. On the other hand, new approaches are characterized by *very free tracking of the central trajectory* where a central trajectory serves only as a compass determining the direction of movement from the interior of a feasible region to the optimal solution at the border. Typical estimate is the number of iterations needed to obtain ε—exact solution.

The *LP problem* in the standard form (4.23) *transformed by the logarithmic barrier function* will have the following form

$$(LP_r)\ \ Min_{x>0}\left\{c^T x - r\sum_{i=1}^{n}\ln(x_i)\ |\ Ax = b,\ r > 0\right\}, \tag{4.25}$$

where $x = (x_1, x_2, ..., x_n)^T$ and its optimal solution is $x^* \in \mathbb{R}^n$. If we denote $x(r)$ the solutions of these problems for a given parameter $r > 0$, then under suitable assumptions it can be shown that the solutions $x(r)$ exist and according to (4.21) it holds that for $r \to 0$, $x(r) \to x^*$ is valid. The Karush- Kuhn-Tucker first-order optimality condition for this system is as follows

$$c - rX^{-1}e = A^T y,\ Ax = b, \tag{4.26}$$

where X is the diagonal matrix with the vector elements x along the diagonal. If we put $z = rX^{-1}e$, we get a system of $2m + n$ linear equations:

$$\begin{aligned} xz &= re \\ A^T y + z &= c\,, \\ Ax &= b. \end{aligned} \tag{4.27}$$

Newton's method is used to solve this system. We solve Newton's step from the system

$$\begin{bmatrix} X & Z & 0 \\ I & 0 & A^T \\ 0 & A & 0 \end{bmatrix} \begin{bmatrix} \Delta z \\ \Delta x \\ \Delta y \end{bmatrix} = \begin{bmatrix} re - xz \\ c - A^T y - z \\ b - Ax \end{bmatrix} \tag{4.28}$$

The system can be further modified using the Gaussian elimination method for a simpler form. We gradually reduce the parameter r during iterations until we reach a suitably selected stopping criterion. To select the stopping criterion, we can use information from the so-called *dual problem* to (LP) problem, i.e., problems (4.24)

$$(DLP)\ Max\left\{ b^T y \mid A^T y + z = c,\ \ z \geq 0,\ y \in \mathbb{R}^n \right\}, \tag{4.29}$$

which gives us a lower estimate of the functional value at the optimal point. See [1, 9, 13]. One of the possible criteria is

$$\frac{\|b - Ax\|}{\max(1, \|b\|)} + \frac{\left\|c - A^T y - z\right\|}{\max(1, \|c\|)} + \frac{\left\|c^T x - b^T y\right\|}{\max\left(1, \left\|c^T x\right\|, \left\|b^T y\right\|\right)} \leq \varepsilon, \tag{4.30}$$

where $\varepsilon > 0$s small enough selected number. Mathematica and MATLAB systems apply this criterion.

In the post-Karmakar period, IPMs for linear programming have been elaborated in detail in several publications [19, 23]. Many new algorithms have been developed, where, in addition to polynomial complexity, the issues of convergence speed, efficient start, termination conditions of calculations, properties of a central trajectory and implementation of the developed algorithms have begun to be investigated. The relevant program became part of the commercial offer. The *simplex method* has been shown to have lost its unique position because IPMs for very large sparse matrix of LP problems are significantly faster than the simplex method. For small and medium-scale problems, we still have a choice between the simplex method and IPM. This choice may depend on the problem to be solved and on the required properties of optimal solutions that the method can provide. While the simplex method provides a basic solution, the interior point method gives a sharply complementary solution, which can sometimes be an advantage. IPM made it possible to understand linear programming as an integral part of convex programming. When applying new IPM procedures to more general convex problems, problems have arisen related to the analysis of the behaviour of the Newton method, i.e., with the so-called Lipschitzianness of the Hessian matrix $\nabla^2 f(x)$, which Nesterov and Nemirovsky solved in 1994 by defining the term *self-condordance*, i.e., a special property of convex objective function. More in [13, 24]. This property of self-concordance made it possible to analyse IPM algorithms for more general convex problems and prove their polynomiality. Interior point methods thus blurred the border between linear and nonlinear programming and created a new border between convex and non-convex programming. They led to the origin of new branches of convex optimization, such as semidefinite programming or its subclass: the so-called second-order cone programming [13]. These have been shown to find application in various fields, such as control theory, design of experiments, structural design, image recognition, or combinatorial optimization.

4.3 Simulated Annealing (SA)

Simulated annealing is a stochastic (randomized) global optimization algorithm. The inspiration for the development of this algorithm was the process of annealing and controlled cooling of metal to achieve the minimum energy of the crystalline structure of the substance. The SA algorithm was developed in 1983 to solve combinatorial and highly nonlinear problems. The advantage of SA is that it possesses properties that allow to find a global optimum and avoid getting stuck in a local optimum. The name of this algorithm comes from the field of metallurgy. Annealing causes atoms in the structure of a substance to become unstable and to move randomly around the original position. Controlled cooling allows atoms to find a position (configuration) with less internal energy than internal energy of atoms in the initial position, thus achieving better properties of a substance. The SA algorithm, in analogy with this physical process, will replace the current solution in each iteration step with a new (quasi) random solution with a certain probability which depends on the change in the value of the objective function and on the global parameter T. In connection with the analogy of the annealing process, the parameter T is called the *system temperature* and during the algorithm we gradually reduce this parameter. We generate new solutions almost randomly if the temperature T is high and with a certain positive probability we also allow solutions in which the value of the objective function increases. Gradually, as we reduce T, we also reduce a probability of generating worse solutions. Positive probability of generating a worse solution prevents the algorithm from getting stuck at the local minimum. The method was independently introduced by S. Kirkpatrick, CD Gelatt and MP Vecchi in 1983 and in 1985 by V. Černý as the adaptation of the Metropolis-Hasting algorithm and the Monte Carlo method generating states of thermodynamic systems [16, 25–27].

The algorithm randomly searches for a new solution while with a positive probability it also accepts those for which the value of the objective function increases. Let $f : \mathbb{R}^n \to \mathbb{R}$ be the objective function that we minimize. Formally, it is possible to describe the structure of the general SA algorithm in the form given in the Table 4.5.

The general scheme of the SA algorithm does not specify the components A, D, T_k and the termination rule. Particular attention must be paid to the selection of these

Table 4.5 Structure of general SA algorithm

Input	Objective function $f : \mathbb{R}^n \rightarrow \mathbb{R}$ Initial point $x^0 \in \mathbb{R}^n$. Let us put $Z_0 := \{x^0\},\ k := 0$
Calculation	(1) We generate a test solution y^{k+1} from potential solutions $D(Z_k)$ (2) We generate number p from distribution $R[0, 1]$ Let us put $x^{k+1} = \begin{cases} y^{k+1} & \text{if } p < A(x^k, y^{k+1}, T_k) \\ x^k & \text{otherwise} \end{cases}$ A—*acceptance function* of values from [0, 1] T_k—*system temperature* in k – th iteration (3) We set $Z_{k+1} = Z_k \cup \{y^{k+1}\}$, set Z_k contains all information about iterated solutions to the point k (4) We set $T_{k+1} = U(Z_{k+1})$, U is a non-negative function, the so-called *cooling scheme* (5) Check the stopping rule, if it does not comply, we set $k = k + 1$ and return to step (1)

components. Modifications of the SA algorithm differs in literature according to their selection and control. These are complex procedures that are mostly part of a software developed for commercial purposes.

We will describe the most common approaches when using the SA algorithm.

The acceptance function A—most often the so-called *metropolitan function* is selected

$$A(x, y, T) = Min\left\{1, e^{-\frac{f(y)-f(x)}{T}}\right\} \tag{4.31}$$

A metropolitan function will always accept steps in which the value of an objective function for a candidate y^{k+1} will decrease with respect to its value at a point x^k. Otherwise, it will accept them with a probability $A(x^k, y^{k+1}, T_k)$ to avoid getting stuck at a local minimum. This probability is controlled by a system temperature T_k, which we gradually reduce and thus reduce the probability of accepting "worse" steps. One of the other options for defining this function is to use the so-called *Barker's criterion*, which may not even accept steps in which an objective function decreases but does not accept these steps with a high probability with decreasing temperature. See [27]. These two acceptance functions are considered essential, as other acceptance functions have been shown to be somewhat equivalent to one of them.

Cooling schedule and generating new solutions

The most important part of defining the SA algorithm is choosing the cooling scheme U and generating a new random solution with distribution D. It is commonly chosen for the cooling scheme

$$U(Z_k) = \beta\big(f(x_k) - f^*\big)^g, \tag{4.32}$$

where $\beta, g > 0$ are constants and f^* is the value of an objective function at an optimal point. It is unknown in most cases, so its estimation $\hat{f}$ is better implemented. The choice of parameters β, g often depends on some properties of a function f, that some SA modifications estimate during the algorithm.

The layout of the newly generated solution can be generally written in the following form

$$y^{k+1} = x^k + \Delta r \theta_k, \tag{4.33}$$

where θ_k is a random vecto satisfying $\|\theta_k\| = 1$ and Δr is a step size. Generating y^{k+1} around x^k in every direction with the same probability brings certain problems. Therefore, information about a local structure of a function f is sometimes used, and steps in certain directions $y^{k+1} = x^k + Q\,u$ are preferred, where Q is a matrix that carries information about a local structure of f and u is $n-$dimensional random vector with a distribution $R\left[-\sqrt{3}, \sqrt{3}\right]^n$. A matrix Q should be modified during the algorithm to carry current information about the local structure. In the so-called *adaptive simulated annealing algorithm*, distribution of a solution is also dependent on a system temperature, the density of the step distribution is chosen according to the following relation

$$g_k(\Delta x) = (2\pi T_k)^{-\frac{n}{2}} e^{-\frac{\|\Delta x\|^2}{2T_k}}, \tag{4.34}$$

For the so-called *Fast Annealing algoritmus* density is given by another relation [28], because for a specific modification of the SA algorithm it is necessary to adjust the cooling scheme as well as the corresponding options.

Termintation rule—due to the nature of the SA algorithm, it is difficult to determine whether the iteration process has reached a global minimum or its approximation with a given accuracy. It is recommended to terminate, if we do not accept a new solution after a certain number of cycles (20–50). The natural rule of termination occurs when iterated solutions begin to differ little from each other, i.e., $|f_i - f_{i-u}| \leq \varepsilon,\ u = 1, \ldots, N_\varepsilon$, for some small $\varepsilon > 0$ and integer N_ε. For example, the Mathematica system applies iterations of this algorithm for multiple initial solutions and a number can be selected. If we denote ax_b^k solution with the smallest functional value obtained in the k - th iteration, then the termination rule applied by this system is as follows

$$\left|f(x_b^k) - f(x_b^{k-1})\right| < \varepsilon \ \text{ and } \ \left|x_b^k - x_b^{k-1}\right| < \varepsilon, \tag{4.35}$$

where $\varepsilon > 0$ is a small enough number.

There is evidence for some specific problems under certain conditions about the convergence of this method, see [28]. The method does not require much functional evaluation of an objective function in the iteration of a step, so it is suitable to use it when the problems with an objective function are difficult to evaluate and, in addition, it is effective in discrete space optimizing.

References

1. Boyd, S., & Vandenberghe, L. (2009). *Convex optimization* (701p.). Cambridge University Press. ISBN 978-0-521-83378-3.
2. Hamala, M. (1972). *Nelineárne programovanie.* Alfa.
3. Hudzovič, P. (2001). *Optimalizácia* (320p.). STU v Bratislave. ISBN 80–227–1598–0.
4. Schrijver, A. (2017). *A course in combinatorial optimization* (221p.). University of Amsterdam.
5. Čermák, L., & Hlavička, R. (2016). *Numerické metódy. Numerical methods* (110p.). CERM. ISBN 978-80-2145437-8.
6. Реклейтис, Г., Рейвиндран, А., & Рэгсдел, К. (1986). *Оптимизация в технике* 1, [Reklaitis, G. V. , Ravindran, A., & Ragsdell, K. M. *Engineering optimization . Methods and applications*] (350p.). Mir, Moskva.
7. Реклейтис, Г., Рейвиндран, А., & Рэгсдел, К. (1986). *Оптимизация в технике* 2, [Reklaitis, G. V. , Ravindran, A., & Ragsdell, K. M. *Engineering optimization. Methods and applications*] (320p.). Mir, Moskva.
8. Bazaraa, M. S., Sherall, H. D., & Shetty, C. M. (1993). *Nonlinear programming. Theory and algorithms* (2nd ed.). Wiley.
9. Dupačová, J., & Lachout, P. (2011). *Úvod do optimalizace* (81p.). Matfyzpress. ISBN 978-80-7378-176-7.
10. Jablonský, J., Fiala, P., & Maňas, M. (1985). *Vícekriteriální optimalizace.* SPN.
11. Taufer, I., Drábek, O., & Javůrek, M. (2010). Metoda simplexů—efektivní nástroj pro řešení optimalizačních úloh. *Řízení a automatizace, XX*(6).
12. Djordievic, S. S. (2020). Some unconstrained optimization methods. In *Applied mathematics.* Intechopen Books. https://doi.org/10.5772/intechopen.83679.
13. Hamala, M., & Trnovská, M. (2012). *Nelineárne programovanie/Nonlinear programming.* Epos. ISBN 978-80-805-7986-9.
14. Antoniou, A., & Lu, W. S. (2007). *Practical optimization. Algorithms and engineering applications* (675p.). Springer Science & Business Media LCC. ISBN-13: 978-0-387-71106-5.
15. Lenstra, J. K., Rinnooy, K., & Schrijver, A. (1991). *History of mathematical programming. A collection of personal reminiscences* (141p.). Elsevier Science Publications.
16. Rao, S. (2009). *Engineering optimization. Theory and practice* (4th ed., 830p.). John Wiley & Sons, Inc.
17. Halická, M. (2004). Dvadsať rokov moderných metód vnútorného bodu. *Pokroky matematiky, fyziky a astronomie, 49*(3), 234–244.
18. Vanderbei, R. J. (2008). *Linear programming: Foundations and extensions* (3rd ed.). Springer.
19. Wright, S. J. (1997). *Primal-dual interior point methods.* SIAM.
20. Bertsimas, D., & Tsitsiklis, J. N. (1997). *Introduction to linear optimization* (p. 608). Athena Scientific.
21. Gill, P. E., Murray, W., Saunders, M. A., Tomlin, J. A., & Wright, M. H. (1986). On projected Newton barrier methods for linear programming and an equivalence to Karmarkar's projective method. *Mathematical Programming, 36*(1986), 183–209. https://doi.org/10.1007/BF02592025.
22. Brunovská, A. (1990) *Malá optimalizácia* (248p.). Alfa, Bratislava. ISBN 80-05-00770-1.
23. Roos, C., Terlaky, T., & Vial, J.-P. (1997). *Theory and algorithms for linear optimization: An interior point approach.* Wiley.
24. Nesterov, Y. E., & Nemirovski, A. S. (1994). *Interior point polynomial algorithms in convex programming.* SIAM Publications.

25. Černý, V. (1985). Thermodynamical approach to the traveling salesman problem: An efficient simulation algorithm. *Journal of Optimization Theory and Applications, 45*(1), 41–51.
26. Kirkpatrick, S., Gelatt, C. D., & Vecchi, M. P. (1983). Optimization by simulated annealing. *Science, 220*(4598), 671–680. https://doi.org/10.1126/science.220.4598.671.
27. Locatelli, M. (2002). Simulated annealing algorithms for continuous global optimization. In: P. M. Pardalos & H. E. Romeijn (Eds.), *Handbook of global optimization. Nonconvex optimization and its applications* (Vol. 62). Springer. https://doi.org/10.1007/978-1-4757-5362-2_6.
28. Avriel, M., Rijckaert, M. J., Wilde, D. J. (1973). *Optimization and design* (2nd ed., p. 489). Cambridge University Press.

Chapter 5
Optimization of Technological Processes

5.1 Technological Processes Control—Optimal Decision Making

The term process refers to transformation of input quantities into output quantities. For example, in the process of anodic aluminium oxidation, transformation occurs under the influence of physical and chemical quantities in the electrolyte (electrolyte temperature, coating time, input voltage, concentration of aluminium cations in the electrolyte, etc.) to the formed anodic layer. In general, we have different types of processes: physical, chemical, biological, technological, manufacturing, sociological, economic, etc. [1–4]. Physical and chemical processes are usually referred to as elementary processes. A set of interconnected physical and chemical processes forms a technological process. A set of technological processes usually forms a production process in which qualitative changes occur to produce a product [2, 3, 5–7]. All processes that are carried out consciously by man must be controlled if we want to achieve the desired goal. *Process control* is the *implementation of a decision* that must be such as to ensure the *selection* of the best, i.e., *optimal solution* from all possible solutions. This is the goal of process optimization—to find the best solution by applying mathematical methods, i.e., using an appropriate formulation of an optimization problem and implementing an appropriate optimization method. However, in the field of scientific and technological activities during control and decision-making processes, decisions are not always made in this way. Very often, a decision is made based on experience and intuition. As stated in [8], in scientific and technological activities control, it is necessary to monitor certain processes and phenomena, analyse them to achieve the optimal course of individual processes and events and determine their sequence to achieve the set goal. This operation is schematically illustrated in Fig. 5.1, according to [9].

This activity (decision making, control) can be performed:

A. Vagaská et al., *Optimization Methods in Mathematical Modeling of Technological Processes*, Mathematical Engineering, https://doi.org/10.1007/978-3-031-35339-0_5

$$\begin{bmatrix}\text{Scientific and}\\ \text{technological}\\ \text{activities}\end{bmatrix} \rightarrow \begin{bmatrix}\text{Evidence of}\\ \text{phenomena}\\ \text{and processes}\end{bmatrix} \rightarrow \begin{bmatrix}\text{Monitoring of mutual relations}\\ \text{– analysis, synthesis}\\ \text{– decision making}\end{bmatrix} \rightarrow \begin{bmatrix}\text{Achieving}\\ \text{the set goal}\\ \text{(optimal solution)}\end{bmatrix}$$

Fig. 5.1 General scheme of scientific and technological activities control

(a) based on a solid logical analysis
(b) based on intuition or practicalism, decisions made accidentally or experimentally,
(c) applying mathematical methods.

Very often decisions are made based on experience. If decisions are based on a solid logical analysis, they can lead to satisfactory results, which, however, cannot be criticized because the optimal value of the results and thus the level of quality of such a decision is not known. We encounter this type of decision-making quite often in the field of *machining technology*, where e.g., *cutting conditions* are chosen based on experience and without a deeper analysis of the machining process and economic factors [10]. Application of mathematical methods to solve a given problem allows us to obtain accurate information about the quality and level of such decisions and solutions. For example, application *of linear programming in the optimization of technological processes* is very popular and very successful as many scientific and technical problems can be described using linear models or it is possible to convert them to these models by appropriate modifications. In machining technology, it is a type of LP problems where limiting conditions of a solution are given by a certain interval and thus have a certain degree of freedom. An extreme case is a problem that does not have constraint conditions (free optimization problem), or a problem with conditions specified so that they lead to only one solution. Usually, there are several options for solving a given problem, and the goal is to find the one that most rationally uses all the options so that the result is optimal in terms of the required criteria.

In accordance with the above, we are talking about *optimal decision-making* which is only possible if we can choose a solution from several variants that are not equivalent. It is an optimal decision-making both at the level of elementary, technological as well as production processes, or processes of strategic importance. In Sect. 2.5, we described the sequence of steps in formulating and solving the optimization problem in the field of engineering optimization, i.e., sequence of steps for optimal decision making:

mathematical model → constraint conditions → objective function → selection of optimization method → software for processing → result verification.

Now, we will take a closer look at the issue of optimal decision-making, i.e., we will be interested in how to reach an optimal decision in the field of optimization of technological processes. We will focus on optimization of the production process, optimization of cutting conditions during machining and optimization of surface treatment processes (anodic aluminium oxidation and galvanizing process).

During the process of optimal decision-making, we are exposed to several problematic issues which in general could be characterized as follows [3]:

1. What is the goal of decision-making? Based on its formulation, we can define the criterion of optimal decision.
2. What conditions must be respected in optimal decision-making? By defining them, we define a feasible region of variants to solve the optimization problem. Conditions and dependencies in an optimized process can also be expressed by complex mathematical relationships which represent a mathematical model of a process.
3. Selection and application of a suitable optimization method.

As stated in [3], characteristics summarized in the three points represent only a minimal range of issues that we must address if we want to reach an optimal decision in production and technological processes optimization.

5.2 Formulation of Optimization Problem

If we want to formulate and solve an optimization problem correctly, we need be clear on the following [3–5]:

1. Optimality criterion and its choice
2. Mathematical model and its compilation
3. Analysis of perturbations
4. Optimization methods: selection, implementation, verification of calculation results.

5.3 Optimality Criterion

Optimality criterion results from the formulation of the goal of an optimal decision-making. According to the formulated goal, the optimality criterion can take the form of a requirement for technical parameters (we require maximum performance, maximum thickness of the formed layer or minimum machine failure, etc.), or economic parameters (we require minimum costs, maximum profit, etc.). If we can write the optimality criterion by a mathematical relation expressing the functional dependence, then we have the so-called criterion function, i.e., the *objective function*. In mathematical expression of the objective function, the following can occur [3, 11]:

1. directly measured input quantities (input parameters that we can influence)
2. indicator expressed by means of input quantities
3. complex indicator expressed by means of input and output quantities
4. functional dependence defined by regression analysis
5. functional dependence determined analytically.

The objective function can also take the form of a simple, *unconditional optimality criterion* when the extremum of the objective function is sought without reference to any other quantities, i.e., it is a free optimization problem, which has already been mentioned in the theoretical part of the monograph. In addition, if we have a conditional optimality criterion, certain constraints are required, for example in GS anodic oxidation process, amount of aluminium in the electrolyte must not exceed 12 $\left[g \cdot l^{-1}\right]$. If we want to control the selected technological process in terms of optimal decision-making, it is necessary to realize that with the same values of input parameters (control variables) it will not be possible to ensure both maximization and minimization of two or more indicators.

5.4 Mathematical Model

If we have several variants of solving a given problem (several combinations of setting input parameters) during a technological process control, it is impractical and difficult to quantify all possible combinations and thus recognize which case is the most appropriate [8, 12, 13]. We have already mentioned in the theoretical part that in such a case it would be an archaic approach [3, 4]. In today's computer age, it is desirable to prioritize a scientific approach in technological processes control, i.e., apply tools and methods of mathematical optimization. Applying appropriate software we can reach an optimal variant systematically, less laboriously and with the required quality level of decisions [14]. In this case, it is necessary to have a mathematical model of a technological or production process. We understand the term mathematical model as relationships between input and output quantities using mathematical relations, i.e., we know the functional dependence between the input factors and the response. In the case of linear programming implementation, a solved problem can be mathematically described by a system of linear equations and inequalities. Mathematically, the description of a system must always include a quantitative criterion based on which we can evaluate the success of selection of input parameters. In some cases, empirical dependencies are already known [5, 9] in other cases it is necessary to formulate them based on statistical analysis of data available from experiments (which we will mention in the case of optimization of the galvanizing process and the process of anodic aluminium oxidation).

Figure 5.2 presents the mathematical model of the process in general. In situation (a) the input variables $x = (x_1, x_2, \ldots, x_n)$ and output variables $y = (y_1, y_2, \ldots, y_m)$ are given without further in-depth analysis. In situation (b), the input variables are divided into those quantities that we use in process control $u = (u_1, u_2, \ldots, u_r)$ and into quantities $z = (z_1, z_2, \ldots, z_k)$, that we do not use in process control. Obviously, $r + k = n$ is valid, where:

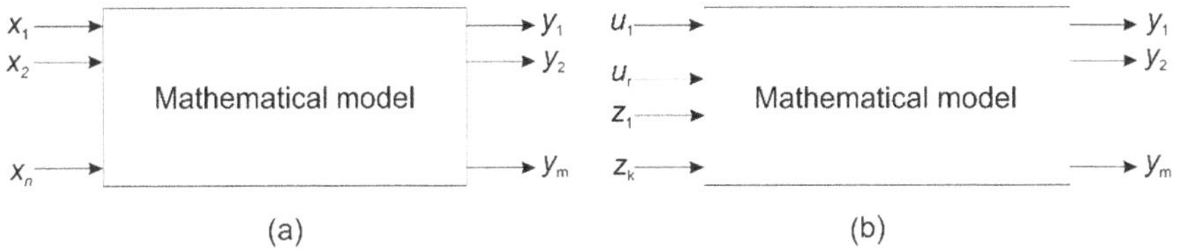

Fig. 5.2 Schematic representation of the process control using the mathematical model

- r—is the number of control (regulatory) variables that we know and want to influence,
- k—is the number of input variables that cannot be influenced during the process,
- n—is the total number of input variables.

For example, in surface treatment process control using anodic aluminium oxidation, the following can be determined as control variables: u_1—amount of sulphuric acid in the electrolyte, u_2—amount of aluminium in the electrolyte, ..., u_r—electrolyte temperature; for uncontrollable variables: z_1– failure of technological equipment (e.g. electrical voltage source), z_2—chemical composition of a material to be plated, ..., z_k—amount of additive elements in sulphuric acid.

Setting of input factors affects the output or response, which in the considered process may be: y_1—the thickness of the formed layer, y_2—the porosity of the formed layer, ..., y_m—the cost of carrying out the anodic aluminium oxidation.

As a result of optimization, we obtain such a combination of values of control quantities $u = (u_1, u_2, \ldots, u_r)$, that after their setting we achieve the required output at the end of the process implementation, i.e., the fulfilment of the defined optimization criterion.

Mathematical models are classified from various points of view in terms of the approach used to compile them. We recognize the following mathematical models [3]:

- *stochastic* (probabilistic, regression)—they are used for modelling processes where input quantities have the character of random variables, but we know their probabilistic characteristics (mean, variance, distribution);
- *deterministic*—their use is possible if we know the behaviour of the process from known physical and chemical laws.

In terms of modelling functional dependencies in the process with an emphasis on time, we distinguish:

- stacionary models
- quasi-stationary
- dynamic

In terms of variables nature describing states in a process, we distinguish:

- discrete models
- continuous models.

The mathematical model of a problem of optimal decision making is usually referred to as an *optimization model*. Regardless of the type, a created optimization model must describe by mathematical means the essential features and characteristics of a real phenomenon or process, which are important for us in terms of our research objective.

5.5 Perturbation Analysis

In the theory of optimal control, perturbation is a change in the state of a process (change in the setting of input quantities x_i) in which calculation of a new optimal mode is required [3]. Perturbation is always caused by changing the setting of input variables and this change can occur slowly or abruptly. The change is also caused by the fact that it is not always possible to stabilize (all) the input non-control quantities $z = (z_1, z_2, \ldots, z_k)$ or to keep them at a set constant value for a long time. Some of them even cannot be measured. For example, during the process of anodic aluminium oxidation, chemical composition of the plating material varies within a certain range, it fluctuates. In this case, we talk about faults of input parameters. Also, control variables $u = (u_1, u_2, \ldots, u_r)$ take values only from a certain interval, not arbitrary interval. In technological processes control, we must always consider certain constraints. In addition to technological constraints, we also consider constraints of an economic nature. Following the above, we can formulate the optimization problem as follows: for a given combination of input values x_i, it is necessary to find such control parameters u_i when all constraints (limiting conditions) are met and the given criterion of optimality is extremized.

5.6 Selection of Optimization Method and Calculation Procedure

We discussed the selection of the suitable optimization method for solving the optimization problem in the theoretical part (Chap. 4). Regarding its implementation, it is appropriate to add that the choice of the optimization method is influenced by the fact how the optimization problem is formulated. If we have a mathematical model of the technological process, which is expressed by a system of equations and inequalities and an optimality criterion is mathematically formulated, then the procedure is obvious. Then, optimization calculation can be performed analytically (rare when solving practical engineering problems) or numerically using suitable software.

Practical engineering problems often require applying of the so-called experimental optimization, i.e., if we cannot describe the properties of systems and processes analytically and we obtained functional dependencies based on statistical analysis of experimentally obtained data. Before the optimization itself, it is advisable to verify the correctness of the programmed methods using the test functions (mentioned in Chap. 3), and only then the criterion function optimization will proceed. In some cases, it is sufficient to use already developed optimization programs, i.e., commercial software.

From the available optimization packages within the software systems, we chose MATLAB to solve the optimization problems mentioned in this monograph. On the webpage introduced in [15], the *Optimization Toolbox Guide* of the company that developed the MATLAB system can be downloaded. This system has built-in optimization programs within a specific toolbox (*optimization toolbox*) for solving constrained and unconstrained optimization problems. It also offers a user a wide range of options to create their own user applications tailored to the problem. MATLAB is a popular software used for solving scientific and engineering problems. Individual optimization programs have been specifically developed for solving problems specific to a given scientific area. *Optimization Toolbox* contains a library of programs, so-called *m-files* also called *functions*, which are used to minimize the criterion function $f(x)$, to solve systems of equations or to approximate data using the least squares method. In Table 5.1 we present an overview of some *m-files* or

Table 5.1 MATLAB programs (functions) for solving optimization problems

Type of optimization problem	Standard form for solutions in MATLAB	The name of the MATLAB program or function to solve a problem
Function of one variable or scala minimization	*To find x minimizing* $f(x)$ *for* $x_1 < x < x_2$	`fminbnd` *(quadratic interpolation method)*
Unlimited minimization of the function of several variables	*To find x minimizing* $f(x)$	`fminunc` *(gradient. m.) or* `fminsearch` *(Nelder-Mead, met. of Simplexes)*
Linear programming	*To find x minimizing* $f^T x$ *subject to* $[A]x \leq b, [A_{eq}]x = b_{eq}$, $l \leq x \leq u$	`linprog` *(Simplex method)*
Quadratic programming	*To find x minimizing* $\frac{1}{2}x^T[H]x + f^T x$ *subject to* $[A]x \leq b, [A_{eq}]x = b_{eq}$, $l \leq x \leq u$	`quadprog`
Minimize the function of multiple variables with constraints	*To find x minimizing* $f(x)$ *subject to* $c(x) \leq 0, c_{eq} = 0$ $[A]x \leq b, [A_{eq}]x = b_{eq}$, $l \leq x \leq u$	`fmincon`
Simulated annealing	*To find x minimizing* $f(x)$ $l \leq x \leq u$	`simulannealbnd`

programs that are offered by the MATLAB optimization toolbox. Their application in optimizing technological processes will be demonstrated in the following chapters.

Using any program or *m-file* from the MATLAB optimization toolbox requires a certain sequence of steps so that the optimization procedure can then be summarized as follows [14]:

- Selection of a suitable program or *m-file* to solve a specific problem.
- Formulation of an optimization problem in the format required by the MATLAB optimization program (function). In general, this includes determining the objective function in the form for "*minimization*" and determining constraints in a specific form "≤", i.e., "*less than or equal to zero*" or for constraints in the form of inequalities we require constant values on the right-hand sides.
- Distinguish between linear and non-linear constraints.
- Identification of lower and upper bounds for input variables.
- Setting/changing the parameters of the optimization algorithm (based on available options).

Then the calculation procedure consists of three steps:

- Step 1 involves writing an *m-file* for the objective function.
- Step 2 is to write an *m-file* for constrains.
- Step 3 requires setting of variable parameters to appropriate values depending on the specifics of the problem and the required output and creating a suitable file to invoke the required MATLAB optimization program (linking the created *m-files* defining the objective function and constraint functions).

Each program or the *m-file* of MATLAB can be implemented in several ways. Details can also be obtained online via the `help` command. For example, the function `fmincon` destined for nonlinear optimization, i.e., minimizing a nonlinear function of several variables with constraints (even with nonlinear ones) can be used in 12 different ways. The differences in the ways of invoking this MATLAB function depend on the data available in the mathematical model of the optimization problem and on the information required in the problem-solving output. If some data are missing in the mathematical model, e.g., we do not have constraints in the form of equations, this must be indicated by using a zero vector in the form „`[]`".

Optimization problems in practical engineering field are characterized by the fact that we are often limited by several constraints. Therefore, it is mainly a solution of optimization problems with constraints. Tasks of this type are generally very complicated and much more difficult to deal with. MATLAB has a powerful tool built into it—„`fmincon ()`", which is helpful in such situations. Demonstrations and numerical illustrations of the use of MATLAB in optimizing the cutting conditions of the turning process and optimizing the technological processes of surface treatment, i.e., the galvanizing process and the anodic aluminium oxidation will be described in the following two chapters of this monograph.

5.7 Demonstration of Optimal Decision Making Using Linear Programming

If we have a choice of several variants and do not want to try all the variants, it is appropriate to use a sequence of steps in the optimal decision:

mathematical model → constraint conditions → objective function → selection of optimization method → software for processing → result verification.

In Chap. 6 we will deal with the problem of cutting conditions optimization in turning using linear programming (after linearization of nonlinear OP by suitable transformation). In this Section we will demonstrate the implementation of the sequence of steps in optimal decision-making using LP application. If the situation and the problem to be solved allows it, it is advantageous to perform a graphical solution of the LP problem which is the simplest way to solve it. However, we can do it only for 2–dimensional optimization problems (number of control variables) or 3–dimensional due to our limited ability in imagination.

For the needs of process optimization, we will briefly outline some issues of linear programming.

The linear programming (LP) problem in general:

$$Min\{f_0(x)\,|\,x \in X, f_i(x) \le 0, i = 1, 2, \ldots, m\} \text{ where } f_0 : X_0 \subseteq \mathbb{R}^n \to \mathbb{R} \quad (5.1)$$

is the *problem of minimizing (or maximizing) of the linear function* subject to the linear constructions. Z viacerých známych tvarov zápisu úlohy LP (primárny, štandardný, kanonický) zrekapitulujeme najpoužívanejšie.

The LP problem in general and detailed form:

$$f_0(\boldsymbol{x}) - c_1x_1 + c_2x_2 + \ldots\ c_nx_n$$

subject to the constraint conditions

$$\begin{aligned} a_{11} \cdot x_1 + a_{12} \cdot x_2 + \ldots + a_{1n} \cdot x_n &\le b_1 \\ a_{21} \cdot x_1 + a_{22} \cdot x_2 + \ldots + a_{2n} \cdot x_n &\le b_2 \\ &\vdots \\ a_{m1} \cdot x_1 + a_{m2} \cdot x_2 + \ldots + a_{mn} \cdot x_n &\le b_n \end{aligned} \quad (5.2)$$

and to the non-negative conditions:

$$x_j \ge 0 \quad j = 1, 2, \ldots, n.$$

The primary form of the LP problem in matrix expression:

$$Min\{\boldsymbol{c}^T\boldsymbol{x} \mid \boldsymbol{A}\boldsymbol{x} \le \boldsymbol{b}, \boldsymbol{x} \ge \boldsymbol{0}\}, \text{ or}$$
$$Min\{\boldsymbol{c}^T\boldsymbol{x} \mid \boldsymbol{A}_1\boldsymbol{x} \le \boldsymbol{b}_1, \boldsymbol{A}_2\boldsymbol{x} = \boldsymbol{b}_2, \boldsymbol{A}_3\boldsymbol{x} \ge \boldsymbol{b}_3, \boldsymbol{x} \ge \boldsymbol{0}\}$$

Matrix notation in standard form of the LP problem:

$$\begin{aligned} f_0(\boldsymbol{x}) = \boldsymbol{c}^T\boldsymbol{x} &\rightarrow \min \\ \boldsymbol{A}\boldsymbol{x} &= \boldsymbol{b} \qquad \text{(SLP)} \\ \boldsymbol{x} &\ge 0 \end{aligned}$$

Optimal solution:

$$\boldsymbol{x}^* \in X \quad \text{if} \quad \forall \boldsymbol{x} \in X : f_0(\boldsymbol{x}^*) \le f_0(\boldsymbol{x}). \tag{5.3}$$

Let us demonstrate the whole procedure of optimal decision-making on a simple illustrative example: a two-dimensional LP problem [16].

Illustrative example:

A factory produces two types of products: A and B. The same equipment is used for assembly and testing of both types of products, the same warehouse for their storage. Specific data on the times required for individual operations, data on the volume of products and the capacity of the operation and warehouse are given in Table 5.2. The profit from the sale of the unit quantity of a product A is 60 EUR and of a product B is 50 EUR. Determine the optimal amount of production of products A and B so that the profit is maximum.

If we think about the solution of the optimization problem outlined in the example, then in terms of the steps sequence in optimal decision-making, we first create a mathematical model, write the constraint conditions and the objective function modelling profit.

Table 5.2 Data overview

	Product		Capacity
	A – x_1	B – x_2	
Assembly [h]	4	10	100
Testing [h]	2	1	22
Data for warehouse [m^3]	1	1	13
Profit from sales [EUR]	60	50	–

$$
\begin{array}{lll}
\text{Constraint conditions :} & \textit{assembly}: & 4x_1 + 10x_2 \leq 100 \\
 & \textit{testing}: & 2x_1 + x_2 \leq 22 \\
 & \textit{warehouse}: & x_1 + x_2 \leq 13 \\
\text{Non-negative conditions :} & & x_1 \geq 0 \\
 & & x_1 \geq 0
\end{array}
$$

Objective function (for profit) : $f_0(\boldsymbol{x}) = f_0(x_1, x_2) = 60x_1 + 50x_2 \rightarrow \max$

It is clear from the example and the created model that the optimality criterion according to which optimization is to be performed, is to achieve the maximum profit. Therefore, the goal is to find such a value of the criterion (objective) function, which will be maximum while observing all prescribed restrictions. This means looking for such a combination of numbers x_1, x_2 from a feasible region that guarantees the maximum value of the objective function $f_0(\boldsymbol{x})$ modelling the profit. We will call this combination of numbers the optimal solution, i.e., $(x_1, x_2) = \boldsymbol{x}^* \in X$. There are many different combinations of production implementations with respect to the number of products. It would be impractical to quantify all possible combinations and thus know the most advantageous production setting. Let us apply mathematical programming, specifically linear programming as the objective function and constraints allows it (they are linear). From the methods of linear programming, we will focus on graphical solution when solving this optimization problem.

Graphical solution of the LP problem—procedure

- Draw constraints (half-planes or constraint lines of constraint conditions).
- Mark the feasible region (convex set)
- Draw a line representing the contour of the objective function for $J = f_0(\boldsymbol{x}) = \mathit{constant}$ and determine the direction of shift to maximum or minimum values.
- Determine the optimal solution located at the extreme point of the feasible region where $f_0(\boldsymbol{x}) = J_{\max}$. We obtain it by moving the line $J = \mathit{const.}$ to the border of feasible region.

Based on the conditions of non-negativity, the feasible region will be in the first quadrant of the Cartesian coordinate system in the plane of the coordinate axes x_1, x_2 representing the change in the number of products of a given type starting at a point [0, 0].

Constraint lines are obtained from constraint conditions:

border line for assembly: $x_2 = 10 - 0.4x_1$ defined by points [0, 10], [10, 6]

border line for testing: $x_2 = 22 - 2x_1$, given by points [11, 0] and [6, 10]

border line for the warehouse: $x_2 = 13 - x_1$, , given by points [0, 13], [13, 0].

Draw the border lines and mark the appropriate half-planes (Fig. 5.3) to obtain a convex set of the feasible region (simplex).

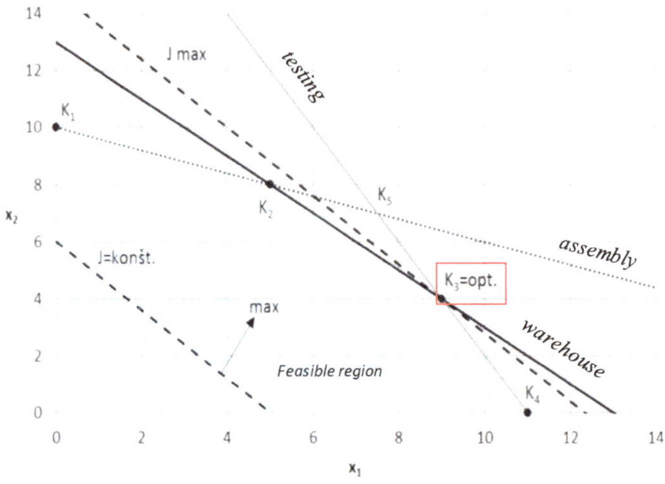

Fig. 5.3 Graphic solution of the example using LP application

To draw the contour of the objective function, let us put $J = const. = 300$ (the smallest common multiple of the coefficients in the objective function equation). From the equation $60x_1 + 50x_2 = 300$ we get two contour points (lines), i.e., for $x_2 = 0$ we get the contour point [5, 0] and for $x_1 = 0$ we get the contour point [0, 6]. By moving the contour line $J = const.$ (the dashed line shown in Fig. 5.3) towards the maximum values, we reach the last extreme point K_3 of the feasible region in which the optimum is located. If we continue to move the contour line, then we get outside the feasible region, the point K_3 is the point of the last contact with the area [14].

Key facts for solving LP problems:

- the feasible region is convex (it is a simplex).
- The optimal solution must be in one of the vertices of the feasible region (simplex vertices) or it can be a whole edge.

By comparing the values of the objective function $J = f_0(x_1, x_2) = 60x_1 + 50x_2$ in the individual vertices K_1, K_2, K_3, K_4 of the feasible region, we can obtain the optimal solution. Let us perform the calculation and find the vertices of the simplex, i.e., vertices on the axes x_1, x_2 and intersections of constraint lines. For $x_1 = 0$ we get the point $K_1(0, 10)$ from the assembly condition; the intersection of the constraint lines for assembly and storage is marked as a point K_2, etc. We clearly write as follows:

$$
\begin{aligned}
&K_1(0, 10)\\
&\quad assembly:\ 4 \cdot x_1 + 10 \cdot x_2 \le 100\\
&\quad border\ line:\ x_2 = 10 - 0.4x_1\\
&J(0, 10) = 60 \cdot 0 + 50 \cdot 10 = 500
\end{aligned}
$$

$$
\begin{aligned}
&K_4(11, 0)\\
&\quad testing:\ 2 \cdot x_1 + x_2 \le 22\\
&border\ line:\ x_2 = 22 - 2x_1\\
&J(11, 0) = 60 \cdot 11 + 50 \cdot 0 = 660
\end{aligned}
$$

$$
assembly \cap warehouse
$$

$$
\left.\begin{aligned} 4 \cdot x_1 + 10 \cdot x_2 &= 100\\ x_1 + \quad x_2 &= 13 \end{aligned}\right\} K_2(5, 8)
$$

$$
J(5, 8) = 60 \cdot 5 + 50 \cdot 8 = 700
$$

$$
assembly \cap testing
$$

$$
\left.\begin{aligned} 4 \cdot x_1 + 10 \cdot x_2 &= 100\\ 2 \cdot x_1 + \quad x_2 &= 22 \end{aligned}\right\} K_5(7.5, 7) \notin X
$$

$$
K_5 - \boldsymbol{infeasible\ solution}
$$

$$
does\ not\ meet\ constraint
$$

$$
testing \cap warehouse
$$

$$
\left.\begin{aligned} 2 \cdot x_1 + x_2 &= 22\\ x_1 + x_2 &= 13 \end{aligned}\right\} K_3(9, 4)\ \boldsymbol{optimum}\ \ for\ warehouse\ x_1 + x_2 \le 13
$$

$$
\boldsymbol{J(9, 4)} = 60 \cdot 9 + 50 \cdot 4 = \mathbf{740}
$$

The optimal solution is a point $\boldsymbol{x}^* = K_3(9, 4)$ because in the sense of the relation (5.3) for $\forall \boldsymbol{x} \in X$ the following holds: $f_0(\boldsymbol{x}^*) \ge f_0(\boldsymbol{x})$, i.e., $f_0(9, 4) \ge f_0(x_1, x_2)$. We have reached this by comparing the values of the objective function at individual points K_1, K_2, K_3, K_4.

We can state that the objective function $J = f_0(\boldsymbol{x})$ modelling the profit reaches its maximum at the point of optimum $\boldsymbol{x}^* = K_3(9, 4)$:

$$
\boldsymbol{J(9, 4)} = f_0(9, 4) = 60 \cdot 9 + 50 \cdot 4 = \mathbf{740}.
$$

Conclusion: The operation will reach a maximum profit value of 740 EUR in the production of 9 units of a product A and at the same time 4 units of a product B.

Discussion about a solution of LP problem:

- one solution (e.g., the presented illustrative example, the solution is one border point, i.e., one vertex of the feasible region),
- infinitely many solutions (the whole edge of the feasible region),
- unlimited solution,
- no solution (if the feasible region is empty).

Considering the above facts for possible cases of solution and key facts for the solution of OP via LP, it is possible in examples with a smaller number of constraints to determine the optimal solution through a sequence of steps:

- calculate the vertices of the feasible region, (each vertex corresponds to a feasible base solution),
- calculate the values of the objective function in individual vertices,
- by comparing these values, we obtain the selection of the optimal solution.

This is also the basis for a numerical search for a solution using the Simplex method, which is based on a systematic transition from one basic solution to another (adjacent) so that the value of the objective function "improves", i.e., increases in case of the maximization problem. A detailed explanation can be found in [8, 16–19]. The goal of solving the LP problem is to obtain the values of variables for maximizing (e.g., profit) or minimizing (e.g., costs, waste) the objective function. If a non-linear objective function or constraints occur, these must be linearized by a suitable transformation. When controlling technological processes, logarithmization to transform into a linear shape is used.

References

1. Jurko, J., Džupon, M., Panda, A., & Zajac, J. (2012). Study influence of plastic deformation a new extra low carbon stainless steels XCr17Ni7MoTiN under the surface finish when drilling. *Advanced Materials Research: AEMT 2012, 538–541*, 1312–1315.
2. Jurko, J., Panda, A., Gajdoš, M., & Zaborowski, T. (2011). Verification of cutting zone machinability during the turning of a new austenitic stainless steel. In: *Advances in Computer Science and Education Applications: International Conference CSE 2011* (pp. 338–345). Springer, 2011.
3. Kostúr, K. (1991). *Optimalizácia procesov* (p. 365). Technická univerzita v Košiciach.
4. Plevný, M., & Žižka, M. (2010). *Modelování a optimalizace v manažerském rozhodování*. Západočeská univerzita v Plzni, Plzeň. 2. vydání. 298 s. ISBN 978-80-7043-933-3.
5. Hudzovič, P. (1990). *Identifikácia a modelovanie*/Identification and modeling (2nd ed., 255p.). Slovenská vysoká škola technická v Bratislave, Bratislava. ISBN 80–227–0213–7.
6. Panda, A., Duplák, J., Jurko, J., & Behún, M. (2013, November 24–25). New experimental expression of durability dependence for ceramic cutting tool. In: *Applied Mechanics and Materials, ICAMM 2012, International Conference on Applied Mechanics and Materials* (Vol. 275–277, pp. 2230–2236), Sanya, China. ISBN 978-303785591-1, ISSN 1660-9336.
7. Panda, A., Duplák, J., Jurko, J., & Pandová, I. (2013). Roller bearings and analytical expression of selected cutting tools durability in machining process of steel 80MoCrV4016. *Applied Mechanics and Materials. Automatic Control and Mechatronic Engineering, 415*, 610–613. ISSN 1660-9336.
8. Kocman, K. (2004). *Speciální technologie. Obrábění* (227p.). 1. vydání. Akademické nakladatelství CERM, s.r.o., Brno, Czech Republic. ISBN 80–214–2562–8.
9. Kocman, K. (2011). *Technologické proces obrábění*. 1. vydání. CERM, Brno. 330 s. ISBN 978–80–7204–722–2.
10. Kocman, K., & Prokop, J. (2003). *Technologie obrábění* (271p.). CERM, Brno, Czech Republic. ISBN 80–214–1996–2.
11. Hudzovič, P. (2001) *Optimalizácia* (320p.). STU v Bratislave, Bratislava, Slovakia. ISBN 80–227–1598–0.
12. Marčuk, G. I. (1987). *Metódy numerické matematiky* (528p.). Academia, nakladatelství ČSAV, Praha, Czechoslovakia.
13. Messac, A. (2015). *Optimization in practice with MATLAB for engineering students and professionals* (496p.). Cambridge University Press. ISBN 978–1–107–10918–6.
14. Rao, S. (2009). *Engineering optimization. Theory and practice* (4th ed., 830p.). John Wiley & Sons, Inc.
15. http://www.mathworks.com/help/toolbox/optim/
16. Rosinová, D., & Dúbravská, M. (2008) *Optimalizácia*. STU v Bratislave, Bratislava. 195 s. ISBN 978-80-227-2795-2.
17. Bertsimas, D., & Tsitsiklis, J. N. (1997). *Introduction to linear optimization* (p. 608). Athena Scientific.

18. Gill, P. E., Murray, W., Saunders, M. A., Tomlin, J. A., & Wright, M. H. (1986). On projected Newton barrier methods for linear programming and an equivalence to Karmarkar's projective method. *Mathematical Programming, 36*(1986), 183–209. https://doi.org/10.1007/BF02592025
19. Vanderbei, R. J. (2008). *Linear programming: Foundations and extensions* (3rd ed.). Springer.

Chapter 6
Application of Mathematical Programming Methods in Optimization of Cutting Conditions in Machining Processes

To produce at low prices, with high quality and, on top of that, in a short time—that's how it has become the motto or hallmark and the main goal for today's production processes. The problem of choosing the optimal cutting parameters in engineering production is therefore still a subject of interest to the professional public and is important for every machining process, since by solving it we can achieve an increase in the quality of machined parts and a reduction in machining costs. In this chapter of the monograph, we will focus on applications of linear and non-linear programming methods in the optimization of cutting conditions during turning. We will be interested in the choice of cutting parameters for numerically controlled machine tools.

6.1 Selection of Optimal Cutting Parameters

In today's industrial world, the issue of increasing the productivity of machine tools and reducing production costs is solved not only through machine tool automation, system rigidity, etc., but also through the optimization of cutting conditions [1, 2]. In case of small series production, it is possible to achieve a higher technological level using numerically controlled machine tools. According to [3–8], numerically controlled machine tools create space in terms of cutting conditions selection:

- ability to program different feed rate values, usually with the smallest programmable unit 0.005 $[\text{mm}\cdot\text{rpm}^{-1}]$ or 0.01$[\text{mm}\cdot\text{min}^{-1}]$,
- possibility to gradually machine several different elements on the machined part with one tool—for example, when turning, it is common to machine a system of cylindrical, conical, face and shaped surfaces with one tool,
- possibility to choose different cutting conditions for machining a set of the above-mentioned elements on one machined semi-finished product. Due to the selected optimality criterion, the resulting tool life will be achieved and observed.

A. Vagaská et al., *Optimization Methods in Mathematical Modeling of Technological Processes*, Mathematical Engineering, https://doi.org/10.1007/978-3-031-35339-0_6

For numerically controlled machines, the fact that numerical control increases the demands on the mechanical load of the toll cutting edge, especially when changing cutting conditions within one machining cycle [9, 10], must also be considered when selecting optimal cutting conditions.

If we plan to implement a machining process and we want to achieve the required results during turning, milling, drilling, threading, or reaming, it is necessary to accurately determine all the cutting parameters values of the process based on working conditions. A necessary condition for the correct determination of cutting conditions is the mathematical formulation of *technological constraints*. These are expressed by relationships in the form of functional dependences on the cutting conditions: *cutting speed* v_c, *feed* f and *depth of cut* a_p. Technological constraints are due to *machine, machine tool* and *workpiece*. This leads to complex calculations. Thus, ways have been and are being sought to simplify the selection of cutting conditions. We know several ways of its implementation. In practice, the most used method of selection is according to the standards of cutting conditions and catalogues of cutting tools manufacturers. Tables of standards are arranged according to the method of machining, work type, tool type, cutting insert material, workpiece material. Nowadays, we can select cutting conditions through the so-called calculators, which are also available in the form of downloadable applications to a computer or mobile phone. In English, they are most often known as *machining calculator*, they are also available in other languages, including Czech and Slovak. As an example, let us mention the Machining Calculator from Sandwik Coromant or the Walter machining calculator, which is also available online and can be used in all operating systems. We also know the method of selection according to nomograms or according to special sliding rulers supplied by tool manufacturers, but in this monograph, we focus our attention on the determination of cutting conditions by calculation and application of mathematical optimization methods.

We will focus specifically on the *optimization of cutting conditions in turning by implementing mathematical programming.* When determining the constraint conditions, we must include in the system of constraints only those that are necessary from the point of view of the optimization calculation.

The basis for optimizing machining parameters is knowledge obtained from the analysis of the cutting process. Statistical models (empirical equations) of the characteristics of the cutting process are the starting point for solving the optimization problem. It must be formulated mathematically so that by its solution we obtain such values of machining parameters using the given production equipment that after their setting we obtain the produced areas with the required quality and at the same time achieve maximum achievement of global production goals at the end of the process.

According to experts in technological machining processes, it is particularly important to achieve the required tool life when determining the cutting conditions regarding the selected optimality criterion.

6.2 Optimal Tool Life

The cutting edge durability is defined by the time during which the tool operates from its sharpening to blunting of the cutting edge. The state of blunting can be assessed from a technological or economic point of view [1, 5, 8, 10–13].

From a technological point of view, the cutting edge is considered blunt if the tool is no longer able to machine the given surface in the required quality.

From an economic point of view, optimal blunting of the cutting edge is evaluated either in terms of maximum service life or it refers to the tool cost share per one piece during machining. From the course of costs and production depending on spindle speed it follows that the correctly selected speed n should be chosen from a certain interval, i.e., $n_1 \leq n \leq n_2$, where

n_1—is a machine spindle speed corresponding to the minimum cost per one component manufacture

n_2—is a machine spindle speed at which the machine is manageable maximum.

The graphical course of costs and production depending on the spindle speed is presented in [1, 4, 5, 8, 12, 14–17].

Optimum tool cutting edge durability for minimum costs can be determined analytically or through optimization using a criterion function expressing costs. Several forms of a cost function are known [1]. When deriving this relationship to achieve economical material removal and thus minimal costs, its analytical solution is based on the total cost $N_{Cj} = N_j$ when machining one *j*-part [18]:

$$N_{C_j} = t_{AS} \cdot \frac{E}{60} + \frac{B}{Q}, \quad \text{resp.} \quad N_j = N_s + N_n + N_v \tag{6.1}$$

where N_{C_j}—is the direct cost for machining of one piece (cost per unit of production) [EUR/pcs]

t_{AS} automatic machine time [min].
E cost of one hour of machine work [EUR / hr].
B the cost of using one cutting wedge [EUR/cutting wedge].
Q number of parts machined between tool changes [pcs].
N_s costs given by the machine.
N_n costs given by the tool.
N_v costs due to consideration of ancillary work.

The relationship (6.1) is modified so that the cost N_j is a function of durability, i.e., based on known relationships for cutting processes we express t_{AS} and Q in the required form:

$$t_{AS} = \frac{L\,p}{n\,f\,a_p} = \frac{L\,\pi\,D\,p}{10^3\,v_c\,a_p\,f} = c_0\,T^{\frac{1}{m}} \tag{6.2}$$

$$Q = \frac{T}{\lambda\, t_{AS}} = \frac{1}{c_0\, \lambda\, T^{\frac{1}{m}-1}}, \text{ where } \lambda = \frac{l}{L} \tag{6.3}$$

while l—is the length of the machined area [mm], L—total length of machine feed. After substitution, we obtain the *cost function* for machining the *j*-th part in the following form:

$$N_{C_j} = c_0\, \frac{E}{60}\, T^{\frac{1}{m}} + B\, \lambda\, c_0\, T^{\frac{1}{m}-1} + N_v, \tag{6.4}$$

Following from mathematical analysis, the minimum (minimum durability) is determined from the necessary condition of the existence of a local extremum, i.e., the first derivative of the function must be equal to zero:

$$\frac{d\, N_{C_j}}{d T} = 0, \tag{6.5}$$

so, after derivation of the (6.4) and modifications we get:

$$\frac{dN_{C_j}}{dT} = c_0\, \frac{E}{60} \cdot \frac{1}{m}\, T^{\frac{1}{m}-1} + B\, \lambda\, c_0 \left(\frac{1}{m} - 1\right) T^{\frac{1}{m}-2} = 0 \;/ \cdot \frac{1}{c_0} \tag{6.6}$$

$$T^{\frac{1}{m}-2} \left[\frac{E}{60} \cdot \frac{1}{m}\, T + B\, \lambda \left(\frac{1}{m} - 1\right)\right] = 0\,. \tag{6.7}$$

If we put $T^{\frac{1}{m}-2} = 0$ in the Eq. (6.6), we get a solution $T = 0$ that has no sense in practice because we do not want zero cutting edge durability. We get the second and suitable solution if we put the expression in square brackets equal to zero and after modifications, we get the relation for optimal durability in terms of minimum costs:

$$\frac{E}{60} \cdot \frac{1}{m}\, T + B\, \lambda \left(\frac{1}{m} - 1\right) = 0 \;\; / \cdot m \tag{6.8}$$

$$T \cdot \frac{E}{60} + B\, \lambda\, (1 - m) = 0 \;\; \Rightarrow \;\; T_{opt1} = (m - 1)\, \frac{60\, B}{E}\, \lambda. \tag{6.9}$$

The value of optimal durability can be determined analytically and in terms of maximum material removal regardless of economy, i.e., when maximum productivity of the machine per unit time is reached. This criterion is based on the total time t_c to machine one piece:

$$t_c = t_{AS} + t_{A11} + \frac{t_{AX}}{Q}, \tag{6.10}$$

where

t_{A11} is the sum of the secondary times [min], $t_{AX}$is time to change and adjust the tool [min].

By analogy with a similar procedure and modifications, the optimum durability for maximum productivity is given by:

$$T_{opt2} = (m - 1)\, t_{AX}\, \lambda. \tag{6.11}$$

As we know, in NC technology, time t_{AX} includes cycle time and tool offset correction time.

6.3 Application of Mathematical Programming to Set Optimal Cutting Parameters

In the theoretical part of the monograph (Sect. 2.2) we defined the *mathematical programming problem* (MP) in the *narrower sense* by the relation (2.4) and by the relation (2.6)

$$Min\{f_0(x)\ | x \in X, f_i(x) \le 0, i \in I,\ h_j(x) = 0,\ j \in J\} \tag{6.2.6}$$

we have defined the *mathematical programming problem* (MP) in the *broader sense,* where I, J are index sets i.e., $I = \{1, 2, \ldots, m\}$ and $J = \{m + 1, m + 2, \ldots, m + r\}$, in case if $x \in X \subset \mathbb{R}^n$ it is a conditional optimization.

When applying mathematical programming to the selection of optimal cutting parameters in turning, it will be a goal to determine the vector of variables

$$x = (x_1,\ x_2,\ \ldots,\ x_n), \tag{6.12}$$

which will ensure the optimum (in terms of both extremization and pareto) of the observed criterion function

$$f_0(x_1,\ x_2,\ \ldots,\ x_n), \tag{6.13}$$

under the fulfilment of conditions (technological constraints) in the form of inequalities, or even equality, i.e.

$$\left.\begin{array}{r} f_1(x_1,\ x_2,\ \ldots,\ x_n) \le 0 \\ \vdots \\ f_m(x_1,\ x_2,\ \ldots,\ x_n) \le 0 \end{array}\right\}, \tag{6.14}$$

and $h_j(x) = 0,\ j \in J$, whereas a non-negative condition is often required for control variables, i.e., $x_i \geq 0$.

If we are looking for optimal values for operating speed n and feed f when optimizing a turning process, then it is obvious to require the non-negative condition to be met. From a mathematical point of view, *multi-criteria goals of mathematical programming* are closer to the real world, but in such a case it is necessary to consider Pareto optimality (non-dominated solutions) [19]. In our case, the solution of the optimal selection of cutting parameters in turning will be the solution of a one-criteria and static problem of mathematical programming.

After the decision to determine the optimal cutting conditions by applying the methods of linear or nonlinear programming, the implementation of the sequence of these steps is as follows:

(a) mathematical formulation of constraint conditions (technological constraints), i.e., analysis of possible selection of cutting conditions
(b) mathematical formulation of the criterion function (based on the mathematical model and formulation of the optimization problem)
(c) selection of optimization method, appropriate software, preparation for optimization procedure (e.g., MATLAB)
(d) solution, computational optimization procedure (if possible, also graphical interpretation of the optimization problem solution)
(e) interpretation of optimization results.

6.4 Constraint Conditions in Machining

During each machining process, certain *technological constraints* occur. When optimizing, we must consider:

- *machine tool constraint conditions* (cutting power, maximum permissible torque, magnitude of the maximum cutting force during machining, strength of the feed mechanism, range of feed of the machine tool, range of cutting speeds of the machine tool),
- *tool constraint conditions* (size of maximum cutting force, size of cutting speed at selected durability),
- *workpiece constraint conditions* (constraints determined by the accuracy of the machined surface, roughness of the machined surface, constraints determined by the depth of cut).

When determining constraint conditions for a specific optimization process, we must include in the system of constraints only those that are necessary in terms of optimization calculation [2–4, 6, 8, 12, 13, 17, 20–24].

We have found some differences in the mathematical formulation of constraint conditions during machining in literature. In [1], we found a tabular summary of constraint conditions during machining, where mathematical formulation of a constraint as well as a constraint meaning according to the fact what constraints the

Table 6.1 Overview of constraint conditions in machining

Constraint cutting ratios given by	Constraint
Machine	Cutting power must be less than or equal to a machine useful power
Machine	Cutting torque must be less than or equal to the torque that can be transmitted by a machine spindle motion mechanism
Machine, tool	Main cutting force must be less than or equal to the force that can be transmitted by a machine's movement mechanism
Machine, tool	Feed force must be less than or equal to the force that a feed mechanism can transmit
Machine	Feed range must be within the working range of a machine
Machine	Cutting speed range must be within the working range of a machine
Tool, workpiece	Cutting speed must correspond to a tool life and material machinability
Tool, workpiece	Cutting conditions must be within the constraints of ensuring production accuracy
Tool, workpiece	Cutting conditions must be within the constraints of ensuring the prescribed surface roughness

cutting ratios, is stated. Let us give an overview without mathematical formulation of constraints (Table 6.1).

There are other constraint conditions that can be considered, and we have not mentioned them. For example, possible vibration during machining, cutting temperature may be limiting in terms of cutting conditions influence, etc. Not all of them always occur in mathematical formulation of the optimization problem. It depends on the user how he models the given technological process and what needs to be considered in the mathematical model. As mentioned in the theoretical part of this monograph, if the model is too complex, the solution to the problem is complicated or even impossible [25]. On the other hand, overly simplified models may not capture reality.

6.4.1 Mathematical Formulation of Constraint Conditions in Turning

When formulating conditions for turning, we will consider:

(a) ***Constraints determined by the range of cutting speed of a tool at selected service life***

If our goal is to *optimize costs* in terms of minimization, then we need to deal with the concept of economic durability of a tool edge. This corresponds to the *economic cutting speed* v_{cT} $[\mathrm{m} \cdot \mathrm{min}^{-1}]$ that is given by the relationship [1]:

$$v_{cT} = \frac{c_v}{T_{opt}^{\frac{1}{m}} a_p^{x_v} f^{y_v}} \; [\mathrm{m} \cdot \mathrm{min}^{-1}] \tag{6.15}$$

The following applies to the cutting speed at a given spindle speed with a given diameter:

$$v_c = \frac{\pi D n}{10^3} \; [\mathrm{m} \cdot \mathrm{min}^{-1}] \tag{6.16}$$

This cutting speed must not exceed the limit value given by the tool sharpness and must therefore apply

$$v_c \le v_{cT}. \tag{6.17}$$

After substitution and modification, we get a constraint condition (C_1) in the form of the inequality:

$$n \cdot f^{y_{v_c}} \le \frac{10^3 \cdot c_{v_c}}{T_{opt}^{\frac{1}{m}} a_p^{x_{v_c}} \pi D} = a_1 \quad (C_1) \tag{6.18*}$$

In which we mark the right side as a_1, as we will use it later in the application of linear programming. T_{opt} is determined by the relation (6.9) or (6.11). Double marking of the relation (6.18*) as (C_1) makes sense with respect to the future graphical solution of cutting conditions optimization.

(b) ***Constraints determined by NC machine tool power***

The relationship between the depth of cut, the feed and the tangential component of the cutting force is given by the experimental relationship:

$$F_c = c_{F_c} a_p^{x_{F_c}} f^{y_{F_c}} \; [\mathrm{N}] \tag{6.19}$$

The dependence of the effective power on the spindle from the limit value of the tangential component of the cutting force $F_{c_{\lim}}$ and from the cutting speed is expressed by the following relation.

$$P_{ef} = \frac{F_{c\lim} v_c}{60 \cdot 10^3} \; [\mathrm{kW}] \tag{6.20}$$

The limit value of the tangential component of the cutting force in terms of effective power is as follows:

$$F_{c\lim} = \frac{6 \cdot 10^7 P_{ef}}{\pi D n} \tag{6.21}$$

The tangential component of the cutting force F_c must not exceed the limit value $F_{C_{\lim}}$

$$F_c \leq F_{C_{\lim}} \tag{6.22}$$

$$c_{F_c}\, a_p^{x_{F_c}}\, f^{y_{F_c}} \leq \frac{6 \cdot 10^7\, P_{ef}}{\pi\, D n} \tag{6.23}$$

Finally, the mathematical formulation of the technological limitation (C_2) determined by machine tool power has the following form:

$$n \cdot f^{y_{F_c}} \leq \frac{6 \cdot 10^7\, P_{ef}}{c_{F_c}\, a_p^{x_{F_c}}\, \pi\, D} = a_2 \quad (C_2) \tag{6.24*}$$

Marking the right side of the inequality by a variable a_2 and the whole relation (6.24*) as (C_2) is again due to the latter application of linear programming (LP) and simplification of the optimization problem writing. We determine the constant c_{F_c}:

$c_{F_c} = 41.84 \cdot (\mathrm{HB})^{0.6077}$ – for grey cast iron.
$c_{F_c} = 176 \cdot (\mathrm{R_m})^{0.359}$ – for steel.

(c) ***Constraint in terms of minimum permissible machine tool productivity (i.e., constraint determined by a production cycle)***

Let us denote Nm the planned machine productivity (in pieces per hour), for its calculation the following relation applies:

$$Nm = \frac{60 \cdot \eta}{t_{AS_1} + \sum t_{A11}} \; [\mathrm{pcs} \cdot \mathrm{hr}^{-1}] \tag{6.25}$$

where

t_{A11} is unit regular time during machine rest,
$\sum t_{A11}$ is the sum of secondary times (parts replacement, clamping, parts disengaging, measurement, etc.) expressed in [min].
η is use of the machine (time utilization of the machine during shifts: downtime, number of simultaneously machined workpieces).

Machine time given by the required productivity:

$$t_{AS_1} = \frac{60 \cdot \eta}{Nm} - \sum t_{A11} \; [\mathrm{min}]^{-1} \tag{6.26}$$

Machine time during machining (turning) of cylindrical surfaces:

$$t_{AS_2} = \frac{L}{n\, f} \cdot \frac{p}{a_p} \; [\mathrm{min}] \tag{6.27}$$

where

L is the total length of a turned cylindrical surface,
p is turning allowance, given in [mm].

If machine productivity is prescribed per unit of time, the actual machine time t_{AS_2} must be less than machine time calculated based on the required productivity:

$$t_{AS_2} \le t_{AS_1} \tag{6.28}$$

$$\frac{L}{n\,f} \cdot \frac{p}{a_p} \le \frac{60\,\eta}{Nm} - \sum t_{A11} \tag{6.29}$$

and after modification we obtain the third constraint condition (C_3):

$$n \cdot f \ge \frac{L}{\left(\frac{60\,\eta}{Nm} - \sum t_{A11}\right)} \cdot \frac{p}{a_p} = a_3 \quad (C_3) \tag{6.30*}$$

(d) ***Constraints determined by cutting parameters of a machine tool (prescribed spindle speed)***

Prescribed spindle speed is another technological constraint in turning. The selected speeds must be greater than minimum spindle speeds. This is expressed by the condition (C_4):

$$n \ge \frac{10^3\, v_{cMIN}}{\pi\, D} = a_4 \quad (C_4) \tag{6.31*}$$

where v_{cMIN} is the minimum permissible cutting speed on diameter D.

The selected speeds must be less than the prescribed maximum spindle speeds. This is expressed by the condition (C_5):

$$n \le \frac{10^3\, v_{cMAX}}{\pi\, D} = a_5 \quad (C_5) \tag{6.32*}$$

where

v_{cMAX} is the maximum permissible cutting speed on the diameter D.

(e) ***Constraints determined by machine tool parameters. Maximum feed is given by the empirical relation***

$$f \le c_f\, r_\varepsilon^{x_\varepsilon}\, a_p^{x_a} = a_6 \quad (C_6) \tag{6.33*}$$

where

r_ε is the radius of curvature of the tool tip given in [mm].
a_p is the cutting width (depth of cut) [mm].

The selected feed rate, dimensionally in [mm·ot^{-1}], must be less than the maximum prescribed feed rate for the given lathe. Specific values c_f, x_ε, x_a for selected materials are given in the tables, e.g. [3, 4, 8, 12, 15, 17, 18].

(f) ***Constraints determined by a system stiffness (limit feed in terms of system stiffness)***

When determining a system stiffness, the force causing component deformation occurs [5], so for the system stiffness we can write as follows:

$$j_s = \frac{F'_p}{\Delta y} \tag{6.34}$$

Based on decomposition of the force $F' = \sqrt{F_c^2 + F_p^2}$ and having indicated the ratio $\frac{F_p}{F_c} = \beta$ we obtain the following relation after its modification

$$F' = \sqrt{F_c^2 + (\beta\, F_c)^2} = F_c\sqrt{1+\beta^2} = F_c\,\lambda_s = c_{F_c}\, a_p^{x_{F_c}}\, f^{y_{F_c}}$$

After substitution we obtain the relation for the system stiffness in the form [1, 10]:

$$j_s = \frac{\lambda_s\, c_{F_c}\, \Delta a_p^{x_{F_c}}\, f^{y_{F_c}}}{\Delta y} \tag{6.35}$$

where

Δa_p is the inaccuracy of the semi-finished product
Δy is inaccuracy after machining, (ratio $\varepsilon = \frac{\Delta a_p}{\Delta y}$ expresses clarification).

The limit displacement in terms of the system stiffness is no longer a problem to obtain by a simple modification from the Eq. (6.35) in the form:

$$f^{y_{F_c}} = \frac{j_s}{\lambda_s\, c_{F_c}\, \frac{\Delta a_p^{x_{F_c}}}{\Delta y}} \quad \Rightarrow \quad f_{\lim} = \left(\frac{j_s}{\lambda_s\, c_{F_c}\, \varepsilon}\right)^{\frac{1}{y_{F_c}}} \tag{6.36}$$

Since the actual feed rate must be less than the limit, we get a_7—i.e., another value of the right side of the constraint condition (C_7) given in [mm · ot^{-1}] with optimization of the turning process for linear programming application:

$$f \le \left(\frac{j_s}{\lambda_s \, c_{F_c} \, \varepsilon} \right)^{\frac{1}{y_{F_c}}} = a_7 \quad (C_7) \tag{6.37*}$$

(g) ***Feed constraint determined by maximum permissible surface roughness***

Cutting conditions must be determined within the constraints of ensuring the prescribed surface roughness. After the machining process realization, the surface roughness (deviation of the machined surface) must be less than or equal to the prescribed surface roughness.

$$R_a \le R_{a\ \max} \tag{6.38}$$

where

R_a[μm] je stredná aritmetická odchýlka obrobeného povrchu
$R_{a\ \max}$ [μm] is the maximum permissible deviation of the machined surface.

The influence of cutting ratios on the microgeometry of the machined surface is the most complex. In addition to the *tool feed* f [$\text{mm} \cdot \text{ot}^{-1}$], the surface roughness in terms of geometry is influenced in particular by the main setting angle χ_r, the secondary setting angle χ_r' and the *radius of curvature of the knife tip* r_ε.

Under the conditions $f < r_\varepsilon$ maximum surface roughness, formed by tool trace after turning (radius of the knife tip), is expressed as follows [8, 17]:

$$R_y \doteq \frac{f^2}{8 r_\varepsilon} \tag{6.39}$$

For the profile of the turned surface, it is possible to obtain a relation for the *mean arithmetic roughness of the surface* R_a for certain machining conditions after modifications in the following form:

$$R_a = 0.26 \, \frac{f^2}{8 r_\varepsilon} \cdot 10^3 \quad [\mu\text{m}] \tag{6.40}$$

Then, for clean turning, the condition (C_8) given in [mm·rpm^{-1}]:

$$f \le \left(\frac{R_a \cdot r_\varepsilon}{32.5} \right)^{\frac{1}{2}} = a_8 \quad (C_8) \tag{6.41*}$$

(h) ***Constraints determined by a system stiffness (prescribed minimum lathe feed)***

There also exists a condition that feed is greater than the minimum allowable value for given cutting and machined material or greater than the minimum feed rate given by technical parameters of the machine tool, expressed in [mm·rpm^{-1}]:

$$f \geq f_{MIN} = a_9 \quad (C_9) \tag{6.42*}$$

Remark: gradual marking for the right-hand sides of the constraint conditions (C_1), ..., (C_9) is to simplify calculations when applying the methods (LP).

6.5 Mathematical Formulation of the Objective Function in Turning

To determine the most advantageous variant of cutting conditions for a turning process, it is necessary to choose the right objective function and its mathematical formulation. We require determination of optimal cutting conditions, i.e., conditions that allow us to machine with minimum production costs and with maximum machine efficiency. Thus, for selection and formulation of the criterion function we will apply the following:

- minimization of production costs per unit of production,
- compliance with maximum productivity while respecting the determined restrictions,
- the aspect of cutting power minimizing while maintaining the required productivity including other constraints.

Several forms of the criterion function designed to *minimize turning costs* are known [1, 3, 4, 12, 15, 17].

For example in publication [1] the total production costs for a single tool turning with constant tool life $T = const.$ is given in the form (we quote in the original designation):

$$N_c = N_{th} + N_{tv} + N_{t_{vym}} + N_{t_{pz}} \tag{6.43}$$

where

N_c are total costs
N_{th} costs determined by main machine time.
N_{tv} costs determined by secondary time.
$N_{t_{vym}}$ costs determined by time of the tool change.
$N_{t_{pz}}$ costs determined by preparation and completion time.

After expressing the individual costs on the right side of the equation, the author states for the total costs as follows:

$$N_c = \underbrace{t_h \left(N_{th} + \frac{N_s}{60} \right)}_{K_1} + \frac{t_h}{T} \underbrace{\left(N_n + t_{vym} \, N_{t_{vym}} \right)}_{K_2} + \underbrace{\left(t_v \, N_{tv} + \frac{t_{pz}}{n} N_{t_{pz}} \right)}_{K_3} \tag{6.44}$$

where

t_h is main machine time
N_s hourly operating costs of the machine.
N_n hourly cost of the tool at a given durability T.
t_{vym} tool change time.
t_v secondary time.
t_{pz} preparation and completion time.
n number of pieces in a batch.

The expressions in parentheses are constants, which, if we gradually denote as K_1, K_2, K_3, then the *cost objective function* takes the following form [1]:

$$N_c = K_1\, t_h + K_2\, \frac{t_h}{T} + K_3 \tag{6.45}$$

This objective function (non-linear) makes it possible to search for optimal values of cutting conditions of single-tool machining with constant tool life $T = const.$ In multi-tool machining, it is necessary to consider different durability of individual tools and changes in cutting speeds when machining individual surfaces with multiple cuts. The durability of the tool T together with the cutting conditions for speed v and feed f always stand out in the criterion functions.

If we follow from the basic law of cutting (extended Taylor's relationship)

$$T = \frac{C_T}{v^m\, f^{y_T}\, a_p^{x_T}}, \tag{6.46}$$

thus, according to this relationship (6.46), the one value of durability T must correspond to any combination of cutting conditions v, f, a_p. This condition must be observed when machining with i–cuts or k–tools. In single tool machining, if the machine feed length $L = const.$ and $a_p = const.$, and the constant K_3 does not contain variables, then we can express the *cost objective function* optimizing the feed f and cutting speed v by formula:

$$N_c = \frac{D_i\, L\, \pi}{10^3} \left(\frac{K_1}{v_i\, f_i} + \frac{K_2'}{v_i^{1-m}\, s_i^{1-y_T}\, c_T} \right), \tag{6.47}$$

The Eq. (6.47) is a *nonlinear objective function* (exponential expression of durability). *Gradient optimization methods* are used to minimize it.

There are other forms of the objective function described in the literature (e.g., Temčin, Goranský, Buda, Koenig, De Piereux, Brown, Rao, etc.). In [1] they are arranged chronologically starting in 1957, where the way in which they were solved for a given period is also mentioned (analytical methods, linear programming, gradient methods, nonlinear programming methods, etc.).

For *linear programming application*, the form of the cost objective function is simplified so that the resulting form can be transformed into linear, e.g., by logarithmization. Therefore, when minimizing production costs, we will follow from direct costs per unit of production [5], which we expressed in Sect. 6.2 by the relationship (6.1)

$$N_{C_j} = t_{AS}\,\frac{E}{60} + \frac{B}{Q}, \tag{6.1}$$

whence after substituting the relation (6.3) for Q and the relation (6.2) for t_{AS}, we get the following

$$N_{C_j} = t_{AS}\left(\frac{E}{60} + \frac{B\,\lambda}{T}\right) = \frac{1}{n\,f}\cdot \underbrace{\frac{L\,p}{a_p}\cdot\left(\frac{E}{60} + \frac{B\,\lambda}{T}\right)}_{K_1} = \frac{K_1}{n\,f}, \tag{6.48}$$

K_1 is a constant for a given case, E are the costs per hour for machine work, B are the costs of the cutting wedge, $\lambda = l/L$ is the ratio of the machined length to the total surface.

The objective function for minimizing production costs in turning (with respect to the unit of production) will be given by the relation:

$$N_{C_j} = \frac{K_1}{n\,f}, \quad (\mathrm{U_N}) \tag{6.49*}$$

The criterion of maximum productivity can be based on unit time

$$t_c = t_{AS} + \frac{t_{AX}}{Q}, \tag{6.50}$$

The objective function for maximum productivity within the given constraints follows from the relation (6.50) after certain modifications by the relation:

$$t_c = \frac{K_2}{n\,f}, \tag{6.51}$$

When formulating the criterion of minimizing cutting power, the starting equation is the relationship for calculating the useful power:

$$P_{ef} = \frac{F_c\,v_c}{60\cdot 10^3}, \tag{6.52}$$

The objective function for cutting power minimizing is defined by the relation:

$$P_{ef} = K_3\,n\,f^{y_{F_c}}. \tag{6.53}$$

In the next chapter we will present the preparation procedure for linear programming application for cutting conditions optimization in turning in terms of production costs minimizing. From this chapter we will use the relationships marked with an asterisk, i.e., for constraint conditions $a_1, a_2, a_3, \ldots, a_9$, gradually the relationships (6.18*), (6.24*), (6.30*)–(6.33*), (6.37*), (6.41*), (6.42*), and for the cost objective function we will use the relationship (6.49*), or conditions sequentially denoted as (C_1), (C_2), ..., (C_9) and the cost objective function as (U_N).

6.6 Preparation for the Optimization Procedure of Cutting Conditions in Turning

We have defined the problem of linear programming (LP) in Sect. 2.3.1 as the problem of minimizing the linear function at linear constraints, where we have introduced several forms, e.g., primary form (P1), (P2) and (P3):

$$Min\left\{\mathbf{c}^T\mathbf{x} \mid \mathbf{Ax} \geq \mathbf{b}, \mathbf{x} \geq \mathbf{0}\right\}, \tag{P3}$$

which is called the standard form of the LP problem. The most used matrix form in standard form of LP problem is as follows

$$\begin{aligned} f_0(\mathbf{x}) = \mathbf{c}^T\mathbf{x} &\to \min \\ \mathbf{Ax} &= \mathbf{b} \\ \mathbf{x} &\geq 0 \end{aligned} \tag{SLP}$$

where $f_0(\mathbf{x}) = c_1 x_1 + c_2 x_2 + \ldots + c_n x_n$ is a linear objective function of several variables, $\mathbf{c} = (c_1, c_2, \ldots c_n)^T$ is a vector of coefficients of the objective function, elements a_{ij} of the matrix $\mathbf{A} \in \mathbb{R}^{m\,x\,n}$ are called constraint coefficients, vector coordinates $\mathbf{b} = (b_1, b_2, \ldots, b_m)^T$ are called coefficients of the right-hand sides of constraints. The matrix $\mathbf{A} \in \mathbb{R}^{m\,x\,n}$ and the vector $\mathbf{b} \in \mathbb{R}^m$ are given, $\mathbf{x} \in \mathbb{R}^n$ is a vector variable.

Methods of solving linear programming problems were presented in Sect. 4.2. To optimize a turning process, we present the application of the Simplex method using MATLAB and the graphical solution of the problem (LP).

In case of turning cutting conditions optimization in sense of their cost minimization, the cost objective function (U_N) or (6.49*) $N_{C_j} = \frac{K_1}{n\,f}$ is the function of two variables (we optimize speed n and feed f), so it is multidimensional optimization. Since both the objective function and the constraints are nonlinear, this is a problem of nonlinear programming (NP), and we could use the methods of nonlinear programming. However, both the objective function and the constraints have a form that allows us to transform the problem (NP) to the problem (LP), i.e., we can linearize the objective function and the constraints by the logarithm.

To optimize a turning process, we write a general mathematical model, the constraints (technological constraints) and the objective function clearly in the following form:

$$\begin{array}{llll}
n \cdot f^{y_{v_c}} & \leq a_1 & (C_1) & (6.18^*) \\
n \cdot f^{y_{F_c}} & \leq a_2 & (C_2) & (6.24^*) \\
n \cdot f & \geq a_3 & (C_3) & (6.30^*) \\
n & \geq a_4 & (C_4) & (6.31^*) \\
n & \leq a_5 & (C_5) & (6.32^*) \\
f & \leq a_6 & (C_6) & (6.33^*) \\
f & \leq a_7 & (C_7) & (6.37^*) \\
f & \leq a_8 & (C_8) & (6.41^*) \\
f & \geq a_9 & (C_9) & (6.42^*) \\
\hline
N_{C_j} = \dfrac{K_1}{n \cdot f} & & (U_N) & (6.49^*)
\end{array}$$

In the next step, it is necessary to perform linearization of the constraints and the objective function, e.g., using common logarithm. Since for the problems (LP) the condition of non-negativity of control variables must be met and for the values from the interval (0,1), e.g. feed $f < 1$, we would get negative numbers after logarithmization, the feed values increase $100 - times$, f will be substituted by $10^2 f$ both in the constraints and the objective function. Having solved the problem, we return to the given substitution. In order the constraint and the objective function remain valid, the whole inequality (6.18*) is multiplied by the expression $10^{2y_{v_c}}$, the inequality (6.24*) is multiplied by the expression $10^{2y_{F_c}}$, Eqs. (6.30*), (6.33*), (6.37*), (6.41*), (6.42*) are multiplied by a number 10^2 and for the objective function, the relation (6.49*) is divided by a number 10^2. Only then we *logarithmize* the constraints and the objective function with a common logarithm. Thus, we get an optimization problem in the form (6.54), where we denote $b_1, b_2, \ldots, b_9$ the logarithmic values of the right sides of the inequalities.

$$\begin{array}{lllll}
\log n & + \; y_{v_c} \; \log 10^2 f & \leq \log(10^{2y_{v_c}} a_1) & = b_1 \\
\log n & + \; y_{F_c} \; \log 10^2 f & \leq \log(10^{2y_{F_c}} a_2) & = b_2 \\
\log n & + \; \log 10^2 f & \geq \log(10^2 a_3) & = b_3 \\
\log n & & \geq \log a_4 & = b_4 \\
\log n & & \leq \log a_5 & = b_5 \\
& \log 10^2 f & \leq \log(10^2 a_6) & = b_6 \\
& \log 10^2 f & \leq \log(10^2 a_7) & = b_7 \\
& \log 10^2 f & \leq \log(10^2 a_8) & = b_8 \\
& \log 10^2 f & \geq \log(10^2 a_9) & = b_9 \\
\hline
\multicolumn{4}{l}{\log N_{C_j} = \log(10^2 K_1) - \log n - \log(10^2 f) \quad \rightarrow \quad \min}
\end{array} \tag{6.54}$$

Having used the substitution

$$\log n = x_1 \tag{6.55}$$

$$\log(10^2 f) = x_2 \tag{6.56}$$

and denotation $y_{v_c} = a_{12}$, $y_{F_c} = a_{22}$, the constraints are modified to the desired form of the LP problem. In addition, if we denote $\log N_{C_j} = f_0(x)$ and a constant $\log(10^2 K_1) = C_0$, then the objective function will have a simplified form:

$$f_0(x) = C_0 - x_1 - x_2. \tag{6.57}$$

The so-called additive objective function $F_0(x)$ to the function $f_0(x)$ will have the following form:

$$F_0(x) = x_1 + x_2. \tag{6.58}$$

For these criterion functions, given that C_0 it is a constant, the following holds:

$$f_0(x_1, x_2) = C_0 - (x_1 + x_2) \rightarrow \min \quad \Leftrightarrow \quad F_0(x_1, x_2) = x_1 + x_2 \rightarrow \max \tag{6.59}$$

In other words: if the sum $(x_1 + x_2)$ or the sum $[\log n + \log(10^2 f)]$ will converge to the maximum value, the costs contained in $f_0(x)$ will converge to the minimum value. After using the introduced substitutions and designations, the problem (6.54) will have the required form of the linear programming optimization problem suitable for application of the Simplex method or graphical solution. *The mathematical model for a turning process optimizing after transformation (linearization)* takes the final form:

$$\begin{array}{llllll}
x_1 & + a_{12} \cdot & x_2 & \leq b_1 & \text{(C1)} & \\
x_1 & + a_{22} \cdot & x_2 & \leq b_2 & \text{(C2)} & \\
x_1 & + & x_2 & \geq b_3 & \text{(C3)} & \\
x_1 & & & \geq b_4 & \text{(C4)} & \\
x_1 & & & \leq b_5 & \text{(C5)} & \text{(OP)} \\
 & & x_2 & \leq b_6 & \text{(C6)} & \\
 & & x_2 & \leq b_7 & \text{(C7)} & \\
 & & x_2 & \leq b_8 & \text{(C8)} & \\
 & & x_2 & \geq b_9 & \text{(C9)} & \\
\hline
\end{array}$$

$$f_0(x) = C_0 - x_1 - x_2 \rightarrow \min, \quad \text{resp.} \quad F_0(x) = x_1 + x_2 \rightarrow \max \tag{6.60}$$

In the form (6.60) of the optimization problem (OP), the constraints are linear, the objective function $f_0(x)$ is linear and also the additive objective function $F_0(x)$. When preparing for the optimization procedure, we calculate the values $b_1, b_2, \ldots, b_9$ on the right sides of the constraints. From the technological constraints (C_1), (C_2), ..., (C_9) we use the relations for $a_1, a_2, \ldots, a_9$. The following will apply to the values $b_1, b_2, \ldots, b_9$ on the right sides of the constraints (OP):

$$b_1 = \log \frac{10^3\, c_{v_c}\, 10^{2y_{v_c}}}{T_{opt}^{\frac{1}{m}}\, a_p^{x_{v_c}}\, \pi\, D} \tag{6.61}$$

$$b_2 = \log \frac{6 \cdot 10^7\, P_{ef} \cdot 10^{2y_{F_c}}}{c_{F_c}\, a_p^{x_{F_c}}\, \pi\, D} \tag{6.62}$$

$$b_3 = \log \frac{L \cdot 10^2}{\left(\frac{60\,\eta}{Nm} - \sum t_{A11}\right)} \cdot \frac{p}{a_p} \tag{6.63}$$

$$b_4 = \log \frac{10^3\, v_{cMIN}}{\pi\, D} \tag{6.64}$$

$$b_5 = \frac{10^3\, v_{cMAX}}{\pi\, D} \tag{6.65}$$

$$b_6 = \log\left(10^2\, c_f\, r_\varepsilon^{x_\varepsilon}\, a_p^{x_a}\right) \tag{6.66}$$

$$b_7 = \log\left[10^2 \left(\frac{j_s}{\lambda_s\, c_{F_c}\, \varepsilon}\right)^{\frac{1}{y_{F_c}}}\right] \tag{6.67}$$

$$b_8 = \log\left[10^2 \left(\frac{R_a\, r_\varepsilon}{32.5}\right)^{\frac{1}{2}}\right] \tag{6.68}$$

$$b_9 = \log\left(10^2\, f_{MIN}\right) \tag{6.69}$$

The goal of the optimization procedure is to find such values x_1, x_2 for which the objective function $f_0(x)$ reaches its minimum or the additive function $F_0(x)$ reaches its maximum on the feasible region defined by the constraints. It is known from the theory of linear programming that the feasible region of the LP problem is convex. If the feasible region is not an empty set, then the objective function assumes its extreme at the extreme point of the convex set. Therefore, if the objective function is a function of two variables, the graphical solution of the optimization problem LP is also approached because it is sufficient to examine the non-negative endpoints of the feasible region.

6.7 Optimization Problem in Turning—Demonstration of Linear Programming Application

Optimal cutting conditions for a turning process can be determined by linear programming methods using appropriate software. The following example shows the possibility of determining the optimal cutting conditions for external longitudinal turning of individual sections of the part (Fig. 6.1) on a numerically controlled lathe applying linear programming methods using Matlab and MS Excel software. The example and values are inspired by [5, 18].

Figure 6.1 presents the diameter and length values for turning *Sections I, II and III*. Let us introduce other necessary values:

Material: 12 051.1 (structural steel C45 according to DIN)
Machinability class: 14b
Tool: cutting insert SK P20
Curvature radius of the knife tip: $r_\varepsilon = 0.8$ [mm]
Time to change and adjust the tool: $t_{AX} = 12.7$ [min]
Spindle machine power: $P_{ef} = 11$ [kW]
Minimum permissible feed: $f_{MIN} = 0.05$ $[\text{mm} \cdot \text{rpm}^{-1}]$
Maximum cutting speed: $v_{C_{MAX}} = 280$ $[\text{m} \cdot \text{min}^{-1}]$
Minimum cutting speed: $v_{C_{MIN}} = 80$ $[\text{m} \cdot \text{min}^{-1}]$
Minimum permissible production: $N_m = 10$ $[\text{pcs} \cdot \text{hr}^{-1}]$
Use of the machine per shift: $\eta = 0.85$
Secondary time (measuring, clamping): $t_{A11} = 0.6$ $[\text{min} \cdot \text{pcs}^{-1}]$
Machining allowance: $p = 4$
Depth of cut: $a_p = 4$ [mm]
Refinement coefficient: $\varepsilon = 4$
Roughness of the machined surface: $R_a = 6.3$ [μm]
System stiffness: $j_s = 4250$ $[\text{N} \cdot \text{mm}^{-1}]$

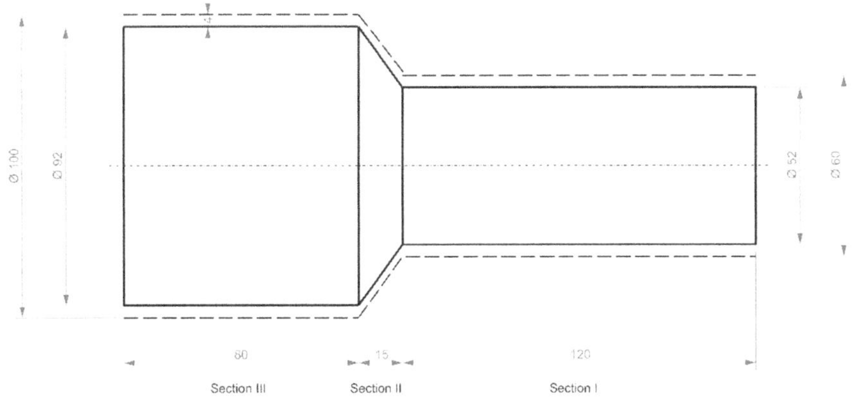

Fig. 6.1 Turned part

Cutting factor: $\lambda \cong 1, \lambda_s = 1.08$

Values of the exponents and constants:

$$\begin{array}{llll} x_{v_c} = 0.11 & x_{F_c} = 1 & c_f = 0.225 & \frac{1}{m} = 0.22 \\ y_{v_c} = 0.25 & y_{F_c} = 0.78 & & x_\varepsilon = 0.83 \\ c_{v_c} = 385 & c_{F_c} = 176 \cdot 500^{0.359} = 1638 & & x_a = 0.338 \end{array}$$

We will use the given values gradually for the calculation $b_1, b_2, \ldots, b_9$ according to the relations (6.61)–(6.69). For example, the values for turning section I we get from (6.61):

$$b_1 = \log(10^{2y_{v_c}} a_1) = \log \frac{10^3 \cdot c_{v_c} \cdot 10^{2y_{v_c}}}{T_{opt}^{\frac{1}{m}} \cdot a_p^{x_{v_c}} \pi D} = \log \frac{10^3 \cdot 385 \cdot 10^{2 \cdot 0.25}}{45.027^{0.22} \cdot 4^{0.11} 60\pi} = 3.38016$$

whereas $T_{opt1} = (m - 1) \cdot t_{AX} \cdot \lambda = \left(\frac{1}{0.22} - 1\right) \cdot 12.7 \cdot 1 = 45.027273$.

We calculate the coefficient b_2 for turning *Section I* according to the relation (6.62):

$$b_2 = \log(10^{2y_{F_c}} a_2) = \log \frac{6 \cdot 10^7 \cdot P_{ef} \cdot 10^{2y_{F_c}}}{c_{F_c} \cdot a_p^{x_{F_c}} \cdot \pi D} = \log \frac{6 \cdot 10^7 \cdot 11 \cdot 10^{2 \cdot 0.78}}{1638 \cdot 4^1 \cdot 60\pi} = 4.28787$$

Analogously, we calculate the values $b_3, \ldots, b_9$ for turning *Section I* by substituting gradually into the relations (6.63)–(6.69). For the following three of them, the following is valid:

$$b_7 = \log\left[10^2\left(\frac{j_s}{\lambda_s c_{F_c} \varepsilon}\right)^{\frac{1}{y_{F_c}}}\right] = \log\left[10^2\left(\frac{4250}{1.08 \cdot 1638 \cdot 4}\right)^{\frac{1}{0.78}}\right] = 1.71614$$

$$b_8 = \log(10^2 a_8) = \log\left[10^2\left(\frac{R_a r_\varepsilon}{32.5}\right)^{\frac{1}{2}}\right] = \log\left[10^2\left(\frac{6.3 \cdot 0.8}{32.5}\right)^{\frac{1}{2}}\right] = 1.59527$$

$$b_9 = \log(10^2 a_9) = \log\left(10^2 \cdot f_{MIN}\right) = \log\left(10^2 \cdot 0.05\right) = 0.69897$$

For turning *Sections II and III*, from the relations (6.61)–(6.69) we once again calculate $b_1, b_2, \ldots, b_9$ for the right sides of the constraints of the LP problem (6.60) and we consider the fact that some coefficients are the same for all sections. Since $a_{12} = y_{v_c} = 0.25$ and $a_{22} = y_{F_c} = 0.78$, then the problem of linear programming (OP) or (6.60) for turning *Sections I, II and III* of the workpiece takes the following form:

LP problem for turning Section I:

$$
\left.
\begin{array}{llll}
x_1 + 0.25\,x_2 \le 3.38016 & (\mathrm{C1}) & & \\
x_1 + 0.78\,x_2 \le 4.28787 & (\mathrm{C2}) & & \\
x_1 + \quad x_2 \ge 3.42597 & (\mathrm{C3}) & & \\
x_1 \qquad \ge 2.62779 & (\mathrm{C4}) & & \\
x_1 \qquad \le 3.17186 & (\mathrm{C5}) & (\mathrm{OP1}) & \\
\qquad x_2 \le 1.47524 & (\mathrm{C6}) & & \\
\qquad x_2 \le 1.71614 & (\mathrm{C7}) & & \\
\qquad x_2 \le 1.59527 & (\mathrm{C8}) & & \\
\qquad x_2 \ge 0.69897 & (\mathrm{C9}) & & \\
\hline
F_0(x) = x_1 + x_2 \quad \rightarrow \quad \max & & &
\end{array}
\right. \tag{6.70}
$$

To analytically calculate the LP problem by the Simplex method, it is necessary to adjust the inequalities and introduce additive (secondary) variables due to the adjustment to the canonical form. This procedure and application of the Simplex method is then implemented for each turning section separately. We will prefer the possibility to use MATLAB software to solve the LP problems for the individual sections.

LP problem for turning Section II:

$$
\left.
\begin{array}{lll}
x_1 + 0.25\,x_2 \le 3.2403 & & \\
x_1 + 0.78\,x_2 \le 4.1480 & & \\
x_1 + \quad x_2 \ge 2.5229 & & \\
x_1 \qquad \ge 2.4879 & & \\
x_1 \qquad \le 3.0320 & (\mathrm{OP2}) & \\
\qquad x_2 \le 1.4752 & & \\
\qquad x_2 \le 1.7161 & & \\
\qquad x_2 \le 1.5953 & & \\
\qquad x_2 \ge 0.6990 & & \\
\hline
F_0(x) = x_1 + x_2 \quad \rightarrow \quad \max & &
\end{array}
\right. \tag{6.71}
$$

LP problem for turning Section III:

$$
\begin{array}{lll}
x_1 + 0.25\,x_2 \le 3.1584 & & \\
x_1 + 0.78\,x_2 \le 4.0660 & & \\
x_1 + \quad\; x_2 \ge 3.2499 & & \\
x_1 \qquad\quad\;\; \ge 2.4059 & & \\
x_1 \qquad\quad\;\; \le 2.9500 & \text{(OP3)} & \\
\qquad\quad\; x_2 \le 1.4752 & & (6.72) \\
\qquad\quad\; x_2 \le 1.7161 & & \\
\qquad\quad\; x_2 \le 1.5953 & & \\
\qquad\quad\; x_2 \ge 0.6990 & & \\
\hline
F_0(x) = x_1 + x_2 \quad \rightarrow \quad \max & &
\end{array}
$$

6.8 Solving the Optimization Problem Using MATLAB

As we have already mentioned, MATLAB has built-in calculation programs (functions) for solving nonlinear (NP) and linear programming (LP) problems [26–32]. In MATLAB, the function for solving LP tasks is predestined as follows:

```
[xo, fo] = linprog(f,A,b,Aeq,beq,lb,ub,x0,options),
```

Which solves optimization problems with constraints in the following form

$$\text{Min } f(\boldsymbol{x}) = \boldsymbol{f}^T\boldsymbol{x} \tag{6.73}$$

subject to constraints:

$$A\,\boldsymbol{x} \le \boldsymbol{b},\quad A_{eq}\,\boldsymbol{x} = \boldsymbol{b}_{eq} \quad \text{and} \quad \boldsymbol{lb} \le \boldsymbol{x} \le \boldsymbol{ub}. \tag{6.74}$$

The coordinates of the vector $\boldsymbol{f}$ are the coefficients of the objective function, A, A_{eq} there are matrices on the left side of the system of constraints in the form of equations or inequalities, the vector $\boldsymbol{lb}$– represents the *lower bound*, $\boldsymbol{ub}$– the *upper bound* for control variables expressed by the vector $\boldsymbol{x} = (x_1, x_2)$. Let us note that without compromising the validity of the relationships, denoting vectors in boldface is not required nowadays. Here, we only used the older vector notation to emphasize the input of input parameters for optimization in MATLAB.

Firstly, MATLAB gives us the optimal values of the vector x from the feasible region, which is called a vector x_0 (or according to the theoretical part of this work $x^* \in K$). As a second output argument, we get the minimized value of the objective function $f(x_0)$ in MATLAB, also known as a minimized *value of* x which is often referred to as `fval`. Some sources state that this function „`linprog()`" works much more efficiently for solving LP problems than the general procedure „`fmincon()`" for optimization tasks with constraints [30, 33].

Program application „`linprog()`" to solve the linear programming problem (LP1), or (6.70) (OÚ1), resp. (6.70) is presented by the following program—script, or m.file „`lp51_AV`". The optimization problem (6.70) is a LP problem in which inequalities also occur in a format other than that accepted by MATLAB. In addition, there are some redundant constraints, this is the upper limit of the variable x_2. From the above, it is obvious that MATLAB solves optimization problems for minimization. Because this is how software programs work, we need to address maximizing of the additive objective function $F_0(x) = x_1 + x_2$ by minimizing the function

$$-F_0(\boldsymbol{x}) = -x_1 - x_2 \quad \rightarrow \quad \min .$$

To create MATLAB's m.file m.file "`lp51_AV`" designed to solve the optimization problem (6.70), we rewrote the problem (OP1) into the following form:

$$\text{Min}\, f(\boldsymbol{x}) = \boldsymbol{f}^T \boldsymbol{x} = [-1 \quad -1]\, [x_1 \quad x_2]^T = -x_1 - x_2 \tag{6.75}$$

$$[A]\, \boldsymbol{x} = \begin{bmatrix} 1 & 0.25 \\ 1 & 0.78 \\ -1 & -1 \end{bmatrix} \begin{bmatrix} x_1 \\ x_2 \end{bmatrix} \begin{matrix} \leq \\ \leq \\ \leq \end{matrix} \begin{bmatrix} 3.38016 \\ 4.28787 \\ -3.42597 \end{bmatrix} = \boldsymbol{b} \tag{6.76}$$

$$\boldsymbol{lb} = \begin{bmatrix} 2.62779 \\ 0.69897 \end{bmatrix} \leq \boldsymbol{x} = \begin{bmatrix} x_1 \\ x_2 \end{bmatrix} \leq \begin{bmatrix} 3.17186 \\ 1.47524 \end{bmatrix} = \boldsymbol{ub} \tag{6.77}$$

To run the optimization procedure for solving the LP problems (6.70)–(6.72) we have created the program "`lp051_AV`":

```
%lp051_AV to solve a Linear Programming problem.
% Max f*x=x(1)+x(2) s.t. Ax <= b, Aeq = beq and lb <= x <= ub
% Min f*x=-x(1)-x(2) s.t. Ax <= b, Aeq = beq and lb <= x <= ub
x0 = [0 0]; %initial point
f = [-1 -1]; %the coefficient vector of the objective function
A = [1 0.25; 1 0.78; -1 -1]; b = [3.38016; 4.28787; -3.42597];
%the inequality constraint Ax <= b
Aeq = []; beq = []; %no equality constraints Aeq*x = beq
lb = [2.62779 0.69897]; ub = [3.17186 1.47524]; %lower/upper
bound lb <= x <= ub
[xo_lp, fo_lp] = linprog(f,A,b,Aeq,beq,lb,ub)
cons_satisfied = [A; Aeq]*xo_lp-[b; beq] %how constraints are
satisfied
f51_AV=inline('-x(1)-x(2)', 'x');
[xo_con, fo_con] = fmincon(f51_AV,x0,A,b,Aeq,beq,lb,ub)
```

The „`linprog()`“ function gives the following ***results of solving the LP problem for Ssection I—problem*** (OP1):

```
>> lp051_AV
Optimization terminated.

xo_lp =
    3.0114
    1.4752

fo_lp =
   -4.4866

cons_satisfied =

   -0.0000
   -0.1258
   -1.0606
```

For comparison, we have built into the program the function „`fmincon()`“ generally used to solve nonlinear programming (NP) problems with constraints. By solving the same problem LP (OP1) or (6.70) by two optimization procedures we obtain the correctness of the results checked:

```
xo_con =     3.0114     1.4752

fo_con =    -4.4866
```

The results as well as more detailed information about the calculation procedure (e.g., number of iterations, algorithm used) can also be worked out by calling a function in the form

```
[x, fval, exitflag, output] =
linprog(f,A,b,Aeq,beq,lb,ub,optimset('Display', 'iter'))

   Optimization terminated.
   x =
       3.0114
       1.4752
   fval =
      -4.4866
   exitflag =
        1
   output =
     struct with fields:
            iterations: 7
             algorithm: 'interior-point-legacy'
          cgiterations: 0
               message: 'Optimization terminated.'
         constrviolation: 0
         firstorderopt: 7.5107e-14
```

We can see that within the built-in program, MATLAB uses the IPM-method, i.e., the *interior point method*, used by the well-known mathematician Karmakar in 1984.

Using "`linprog()`" and calling the function (m.file) "`lp051_AV`" we get the optimal values of the vector $\boldsymbol{x} = (x_1,\ x_2)$.

Optimal solution of the problem (OP1) ***for transformed variable***

$$\mathbf{x}^* = \mathtt{x0\_\ lp} = (3.0114;\ \ 1.4752),$$

i.e., optimal values of the vector $\boldsymbol{x}^*(x_1, x_2) \in X$, in which the additive function $F_0(\boldsymbol{x})$ reaches its maximum

$$\begin{matrix} F_0(\boldsymbol{x}) = x_1 + x_2 \quad \rightarrow \quad \max \\ F_0(\boldsymbol{x}^*) = F_{0\,MAX} = 4.4866 \end{matrix} \quad \text{whereas} \quad \begin{matrix} -F_0(\boldsymbol{x}) = -x_1 - x_2 \quad \rightarrow \quad \min \\ -F_0(\boldsymbol{x}^*) = -4.4866 \end{matrix}$$

Optimal values of cutting conditions (speed, feed) for Section I

Upon return to substitution (6.55) $\log n = x_1$ and substitution (6.56) $\log(10^2 f) = x_2$, i.e., by the inverse transformation we can determine the optimal values for spindle speed and tool feed during turning *Section I*:

$$\log n = x_1 = 3.0114 \quad \Rightarrow \quad n_{opt\ \mathrm{I.}} = 10^{3.0114} = 1026.59\ [\mathrm{rpm}]$$

$$\log(10^2 f) = x_2 = 1.4752 \quad \Rightarrow \quad f_{opt\ \mathrm{I.}} = 0.01 \cdot 10^{1.4752} = 0.298675\ [\mathrm{mm} \cdot \mathrm{rpm}^{-1}]$$

We conceived the problem LP (6.70) in such a way that by maximizing the additive function $F_0(x) = x_1 + x_2 \quad \rightarrow \quad \max$ we minimized the original cost objective

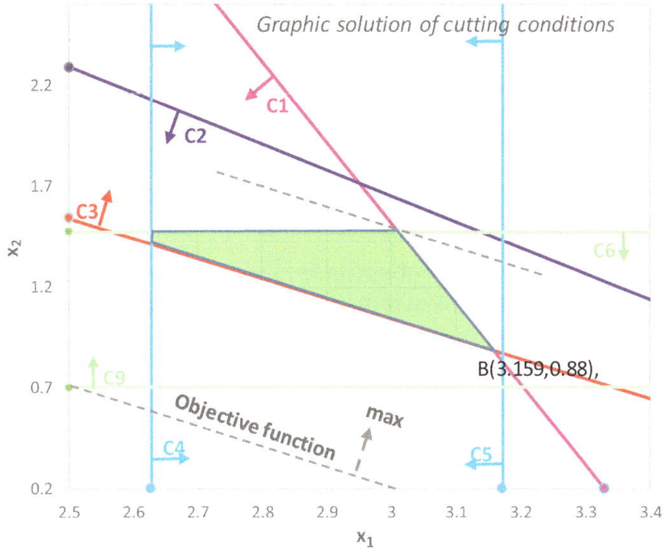

Fig. 6.2 Graphic solution for Section I in turning using LP

function $f_0(x) = C_0 - (x_1 + x_2) \quad \rightarrow$ min, where $f_0(x) = \log N_{C_j}$ and $C_0 = \log(10^2 K_1)$. From solving the problem in MATLAB we have:

```
fo_lp = -4.4866.
```

In our case $F_{0MAX} = 4.4866$. We take a positive value because we had to solve the maximization through the minimization of the function $-F_0(x) = -x_1 - x_2 \quad \rightarrow$ min, because that is how SW programs work.

If we return to the next substitution and marking, we get for costs:

$$\log N_{C_j} = C_0 - F_{0MAX} = \log(10^2 K_1) - 4.4866 \tag{6.78}$$

and then a ***functional dependence for the cost function in Section I***:

$$N_{C_j} = 10^2 \cdot K_1 \cdot 10^{-4.4866} = K_1 \cdot 10^{-2.4866}. \tag{6.79}$$

Figure 6.2 presents a graphical solution of the problem (OP1).

It is the graphical solution of the linear programming problem (OP1) for *Section I*, *i.e.*, the problem (6.70). Figure 6.1 graphically represents the constraints (C_1)–(C_6), (C_9), the criterion function and the feasible region. Constraints (C_7) and (C_8) proved to be redundant. As presented in the illustrative example (introduced in Sect. 5.7), we obtain the optimal solution by moving the criterion function towards the maximum values.

Results of solving the LP problem for Section II—problem (OP2):

In a similar way as for *Section I* of the turned part, we obtained the optimal values x_1, x_2 of the optimization problem (OP2) for *Section II* of the turned part by calling the function „`linprog()`" or „`fmincon()`". We worked again with a modified additive criteria function $-F_0(x) = -x_1 - x_2 \rightarrow$ min suitable for MATLAB: We have created a script „`lp052_AV.m`", thanks to which we have the following results for *Section II*:

```
%lp052_AV to solve a Linear Programming problem.
f = [-1 -1]; %the coefficient vector of the objective function
A = [1 0.25; 1 0.78; -1 -1]; b = [3.2403; 4.1480; -2.5229]; %the
inequality constraint Ax <= b
Aeq = []; beq = []; %no equality constraints Aeq*x = beq
lb = [2.4879 0.6990]; ub = [3.0320 1.4752]; %lower/upper bound lb
<= x <= ub
[x, fval, exitflag, output] = linprog(f,A,b,Aeq,beq,lb,ub)
cons_satisfied = [A; Aeq]*xo_lp-[b; beq] %how constraints are
satisfied
```

```
>> lp052_AV

[xo_lp, fo_lp] = linprog(f,A,b,Aeq,beq,lb,ub)

Optimization terminated.

   xo_lp =
       2.8715
       1.4752
   fo_lp =
      -4.3467
   cons_satisfied =
      -0.0000
      -0.1258
      -0.3043
```

Optimal solution of the problem (OP2) ***for transformed variables***:

$$\mathbf{x}^* = \texttt{x0\_ lp} = (2.8715;\ \ 1.4752),$$

i.e., optimal values of the vector $\boldsymbol{x}^*(x_1, x_2) \in X$, in which the additive function $F_0(\boldsymbol{x})$ reaches its maximum:

$$\begin{matrix} F_0(\boldsymbol{x}) = x_1 + x_2 \quad \rightarrow \quad \max \\ F_0(\boldsymbol{x}^*) = F_{0\,MAX} = 4.3467 \end{matrix} \quad \text{whereas} \quad \begin{matrix} -F_0(\boldsymbol{x}) = -x_1 - x_2 \quad \rightarrow \quad \min \\ -F_0(\boldsymbol{x}^*) = -4.3467 \end{matrix}$$

Optimal values of cutting conditions (speed, feed) for section II

After returning to the substitutions (6.55) $\log n = x_1$ and (6.56) $\log(10^2 f) = x_2$, i.e., we can determine the optimal values for spindle speed and tool feed during turning in *Section II* by reverse transformation:

$$\log n = x_1 = 2.8715 \quad \Rightarrow \quad n_{opt\ \text{II.}} = 10^{2.8715} = 743.875 \cong 744\ [\text{rpm}]$$

$$\log(10^2 f) = x_2 = 1.4752 \quad \Rightarrow \quad f_{opt\ \text{II.}} = 0.01 \cdot 10^{1.4752} = 0.298675\ [\text{mm} \cdot \text{rpm}^{-1}]$$

From the solution of (OP2) in MATLAB we have `fo_lp = -4.3467`, so the calculated maximum value of the additive criterion function is $F_{0MAX} = 4.3467$, where we get:

$$F_{0MAX} = 4.3467 \Rightarrow \log N_{C_j} = C_0 - F_{0MAX} = \log(10^2 K_1) - 4.3467$$

For the cost function of Section II, the following will apply:

$$N_{C_j} - 10^2 \cdot K_1 \cdot 10^{-4.3467} - K_1 \cdot 10^{-2.3467} \tag{6.80}$$

Graphic solution for this section will be implemented analogously as for *Section I*.

<u>***Results of solving the LP problem for Section III—problem*** (OP3):</u>

To run the optimization procedure for solving the problem (OP3), we have again modified the additive criterion function to the form $-F_0(x) = -x_1 - x_2 \quad \rightarrow \quad \min$ that is suitable for MATLAB. The m.file „`lp053_AV.m`", was created thanks to which we obtained the results for *Section III*:

```
>> lp053_AV
Optimization terminated.
x =
    2.7896
    1.4752
fval =

   -4.2648
exitflag =
     1
output =
  struct with fields:
         iterations: 1
    constrvioaltion: 0
            message: 'Optimal solution found.'
          algorithm: 'dual-simplex'
      firstorderopt: 4.4409e-16

cons_satisfied =

   -0.0819
    0.0438
   -1.0968
```

Optimal solution of the problem (OP3) ***for transformed variables***:

$$\mathbf{x}^* = \texttt{x0\_ lp} = (2.7896;\ \ 1.4752),$$

i.e., optimal values of the vector $\boldsymbol{x}^*(x_1, x_2) \in X$ in which the additive objective function $F_0(\boldsymbol{x})$ reaches its maximum:

$$\begin{array}{lcl} F_0(\boldsymbol{x}) = x_1 + x_2 \quad \rightarrow \quad \max & \text{whereas} & -F_0(\boldsymbol{x}) = -x_1 - x_2 \quad \rightarrow \quad \min \\ F_0(\boldsymbol{x}^*) = F_{0MAX} = 4.2648 & & -F_0(\boldsymbol{x}^*) = -4.2648 \end{array}$$

Optimal values of cutting conditions (speed, feed) for Section III

After returning to the substitutions (6.55) $\log n = x_1$ and (6.56) $\log(10^2 f) = x_2$, we obtain optimal values for spindle speed and tool feed in *Section III* in the following form:

$$\log n = x_1 = 2.7896 \quad \Rightarrow \quad n_{opt\ \text{III.}} = 10^{2.7896} = 616.027 \cong 616\ [\text{rpm}]$$

$$\log(10^2 f) = x_2 = 1.4752 \quad \Rightarrow \quad f_{opt\ \text{III.}} = 0,01 \cdot 10^{1.4752} = 0.298675\ [\text{mm} \cdot \text{rpm}^{-1}]$$

From the solution of (OP3) in MATLAB we have `fo_lp = -4.2648`, so the calculated maximum value of the additive criterion function is $F_{0MAX} = 4.2648$, where we get:

$$F_{0MAX} = 4.2648 \Rightarrow \log N_{C_j} = C_0 - F_{0MAX} = \log(10^2 K_1) - 4.2648$$

For the cost objective function in section III, the following will apply

$$N_{C_j} = 10^2 \cdot K_1 \cdot 10^{-4.2648} = K_1 \cdot 10^{-2.2648} \tag{6.81}$$

Having applied linear programming, we have achieved the goal of optimization: determination of optimal cutting conditions in turning with respect to the selected optimality criterion, i.e., cost minimization criterion:

$$N_{C_j} = t_{AS}\left(\frac{E}{60} + \frac{B\lambda}{T}\right) = \frac{1}{n\,f} \cdot \underbrace{\frac{L\,p}{a_p} \cdot \left(\frac{E}{60} + \frac{B\lambda}{T}\right)}_{K_1} = \frac{K_1}{n\,f} \quad \rightarrow \quad \min$$

If we want to achieve minimum costs when turning individual sections of a part, it is recommended to set the ***optimum speed*** and ***optimal feed*** on the lathe, i.e., calculated values:

Section I	*Section II*	*Section III*
$n_{opt\ \mathrm{I.}} = 1026\,[\mathrm{rpm}]$	$n_{opt\ \mathrm{II.}} = 744\,[\mathrm{rpm}]$	$n_{opt\ \mathrm{III.}} = 616\,[\mathrm{rpm}]$
$f_{opt\ \mathrm{I.}} = 0.298\,[\mathrm{mm/rpm}]$	$f_{opt\ \mathrm{II.}} = 0.298\,[\mathrm{mm/rpm}]$	$f_{opt\ \mathrm{III.}} = 0.298[\mathrm{mm/rpm}]$
$N_{C_j} = K_1 \cdot 10^{-2.4866}$	$N_{C_j} = K_1 \cdot 10^{-2.3467}$	$N_{C_j} = K_1 \cdot 10^{-2.2648}$

In practical conditions, when selecting specific speeds on the machine, the actual speeds selected may differ from the optimum ones. Then the optimal durability of the tool changes to the so-called the resulting durability, which can be calculated (regarding time or number of pieces).

Analogously, it is possible to implement the optimization procedure for:

- *criterion of maximum production (per unit time)*
- *criterion of minimizing cutting power.*

References

1. Kažimír, I., & Beňo, J. (1989). *Teória obrábania. Návody na cvičenia* (280p.). Alfa.
2. Loladze, A. T. (1989). *Základy optimalizácie strojárskej technológie* (216p.). Alfa. ISBN 80–05–00083–9.
3. Buda, J., & Békés, J. (1977) *Teoretické základy obrábania kovov*. Alfa.
4. Buda, J., Souček, J., & Vasilko, K. (1983). *Teória obrábania*. Alfa.
5. Kocman, K. (2004). *Speciální technologie. Obrábění* (227p.). 1. vydání. Akademické nakladatelství CERM, s.r.o. ISBN 80–214–2562–8.

6. Mattson, M. W. (2010). *CNC programming. Principles and applications* (1st ed., 432p.). Delmar, Cengage Learning, Inc. ISBN 10: 0766818888.
7. Michalík, P., Dupläk, J., & Kormoš, M. (2015) *Číslicovo riadené obrábacie stroje a ich programovanie* (205p.). FVT TU v Košiciach so sídlom v Prešove. ISBN 978-80-553-1837-0.
8. Vasilko, K. (2009). *Teória a prax trieskového obrábania* (532p.). FVT TU v Košiciach. ISBN 978-80-553-0152-5.
9. Antoniou, A., & Lu, W. S. (2007). *Practical optimization. algorithms and engineering applications* (675p.). Springer Science & Business Media LCC. ISBN-13: 978-0-387-71106-5.
10. Kocman, K., & Prokop, J. (2003). *Technologie obrábění* (271p.). CERM. ISBN 80–214–1996–2.
11. Fulemová, J., & Řehoř, J. (2015, December). Influence of form factor of the cutting edge on tool life during finishing milling. *Procedia Engineering, 100*, 682–688. Elsevier LTd. https://doi.org/10.1016/j.proeng.2015.01.420.
12. Mádl, J. (1989). *Teórie obrábění*. 1. vyd. Praha: ČVUT.
13. Wang, J., Kuriyagawa, T., Wei, X. P., & Guo, D. M. (2002). Optimization of cutting conditions for single pass turning operations using a deterministic approach. *International Journal of Machine Tools and Manufacture, 42*(9), 1023–1033. ISSN 0890-6955, https://doi.org/10.1016/S0890-6955(02)00037-8.
14. Panda, A., Vasilko, K., & Dupläk, J. (2011, March 19–20). Evaluation of T–v_c dependence in standard ISO 3685 for selected cutting materials. In *Asia–Pacific conference on wearable computing systems, APWCS2011* (Vol. 2, pp. 129–132). IEEE. ISBN 978-1-4244-9870-3.
15. Přikryl, Z., & Musílková, R. (1975). *Teorie obrábění*. SNTL.
16. Serra, R., Chibane, H., & Duchosal, A. (2018). Multi-objective optimization of cutting parameters for Turning AISI 52100 hardened steel. *The International Journal of Advanced Manufacturing Technology, 99*, 2025–2034. Springer-Verlag London Ltd.
17. Valíček, J., Harničárová, M., Panda, A., Hlavatý, I., Kušnerová, M., Tozan, H., Yagimli, M., & Václavík, V. (2016). Mechanism of creating the topography of an abrasive water jet cut surface. In A. Öchsner & H. Altenbach (Eds.), *Machining, joining and modifications of advanced materials. Advanced structured materials* (Vol. 61). Springer. https://doi.org/10.1007/978-981-10-1082-8_12.
18. Kocman, K. (2011). *Technologické proces obrábění*. 1. vydání. CERM. 330 s. ISBN 978–80–7204–722–2.
19. Jablonský, J., Fiala, P., & Maňas, M. (1985). *Vícekriteriální optimalizace*. SPN.
20. Badida, M., Gombár, M., Mašlejová, A., Sobotová, L., Kmec, J., & Vagaská, A. (2015). Evaluation of zinc coating quality by statistical methods. *Przemysl Chemiczny, 94*(12), 2146–2149. https://doi.org/10.15199/62.2015.12.11.
21. Hudzovič, P. (1990). *Identifikácia a modelovanie/Identification and modeling* (2nd ed., 255p.). Slovenská vysoká škola technická v Bratislave. ISBN 80–227–0213–7.
22. Michal, P., Vagaská, A., Gombár, M., Kmec, J., Spišák, E., & Kučerka, D. (2015). Usage of neural network to predict aluminium oxide layer thickness. *The Scientific World Journal*, Article number 253568, 10p. https://doi.org/10.1155/2015/253568.
23. Panda, A., Jurko, J., Džupon, M., & Pandová, I. (2011). Optimalization of heat treatment bearings rings with goal to eliminate deformation of material. In *Chemické listy*, Material in Engineering Practice 2011 (Vol. 105, pp. 459–461). Special issue, Association of Czech chemical society Praha, HF TU Košice, ISSN 0009-2770.
24. Vagaská, A., Gombár, M., & Straka, Ľ. (2022). Selected mathematical optimization methods for solving problems of engineering practice. *Energies, 15*(6), 2205. https://doi.org/10.3390/en15062205
25. Rosinová, D., & Dúbravská, M. (2008). *Optimalizácia*. STU v Bratislave. 195 s. ISBN 978-80-227-2795-2.
26. Dušek, F. (2000). *Matlab a Simulink* (147p.). Univerzita Pardubice. ISBN 80-7194-273-1.
27. Hrehová, S. (2019, October 8–10). Possibilities of data analysis using data model. In *Proceedings of the 4th EAI international conference on Management of Manufacturing System (MMS 2019)*, Krynica Zdroj, Poland. EAI/Springer Innovations in Communication and Computing; Springer. https://doi.org/10.1007/978-3-030-34272-2_20.

28. Kmeť, T., Vozár, M., Kmeťová, M. (2012). *MATLAB a vizualizácia numerických a optimalizačných metód* (191p.). Fakulta prírodných vied UKF v Nitre. ISBN 978-80-558-0114-8.
29. Moré, J. J., & Wright, S. J. (1993). *Optimization software guide*. Society of Industrial and Applied Mathematics.
30. Rao, S. (2009). *Engineering optimization. Theory and practice* (4th ed., 830 p.). John Wiley & Sons, Inc..
31. Volauf, P. (2005). *Numerické a štatistické výpočty v Matlabe*. STU v Bratislave. ISBN 80-227-2259-6.
32. Zaplatílek, K., & Doňar, B. (2008). *MATLAB—tvorba uživatelských aplikací* (216p.). BEN-technická literatura. ISBN 978-80-7300-133-9.
33. Yang, W. Y., Cao, W., Chung, T. S., & Morris, J. (2005). *Applied numerical methods using MATLAB* (p. 509.). John Wiley & Sons, Inc.

Chapter 7
Application of Nonlinear Programming Methods in Optimization of Surface Treatment Processes

7.1 Application of Nonlinear Programming to Optimize the Zincing Process

Mathematical optimization has become an important tool in many areas and the list of applications is constantly expanding. In most applications, mathematical optimization assists with decision-making. It is helpful for a designer of a certain object or system, for a system operator who oversees the process, checks the results and, if necessary, adjusts the problem model or approach to the solution. Such a decision-maker needs to direct all activities related to the optimization problem, such as buying or selling, to achieve optimal profits. Optimization problems proved to be significant both in engineering facilities design and their realization as well as in a design stage of technological processes and their operation.

In this chapter we will deal with mathematical optimization in mechanical engineering field focusing on a special role of convex and nonlinear programming in solving optimization problems in connection with the optimization of technological processes of surface treatment. In metal surface treatment processes, the process duration (plating time) is one of the most important parameters that determine the efficiency of the whole process. If we manage to minimize the time necessary to create a coated layer with the required thickness while maintaining the desired quality by setting the control variables (input factors), it is possible to maximize the economic profit. We will introduce the application of nonlinear programming to optimize the galvanizing process and anodic aluminium oxidation (anodizing) using MATLAB software tools. We will discuss the influence of input factors on that of output.

7.1.1 Galvanizing Process Analysis

Technological processes of surface treatment belong to complex multifactor systems. They represent complex nonlinear processes with the action of several technological,

A. Vagaská et al., *Optimization Methods in Mathematical Modeling of Technological Processes*, Mathematical Engineering, https://doi.org/10.1007/978-3-031-35339-0_7

physical, chemical, and material influences and their mutual interactions. Therefore, the analysis of these processes by classical methods appears to be inefficient and often leads to incorrect conclusions. In their analysis, i.e., research, testing, comparison, and synthesis as well as in optimization and forecasting we start from the determination of relationships between input and output quantities. It is important to identify whether certain factors (input variables) affect the monitored variable (response). This is followed by the goal of finding such levels of factors that the optimum (maximum, minimum) of the observed quantity is achieved [1–5]. It is more appropriate to prefer an experimental and statistical approach over a determinative one when solving practical problems. In the experimental-statistical method of analysis of surface treatment processes, we perform analysis and synthesis in conditions of incomplete information. Although the nature of the investigated process is not completely known, we supplement this incomplete information with experiments and known data to determine optimal conditions. The widespread use of various experimental methods is the result of incomplete information and at the same time it is a part of the constant improvement of old as well as creation of new objects, processes, and procedures [1, 6–10]. Applying a design of experiments method in research activities (analysis, modelling and identification) makes it possible to acquire new basic knowledge on technological processes and methods of their control. The result of such activities should be a model that allows to predict the values of the investigated optimization parameter (dependent variable) for different points of a factor space of independent variables. Planning of experiments and analysis of the obtained results are important stages in revealing the essence and course of the investigated process.

Multifactor and nonlinear systems also include the process of electro-galvanizing, where with the right selection (setting) of technological factors it is possible to create a protective layer of material (part) that will have the required thickness and properties and will meet predefined criteria (e.g., corrosion resistance). Zinc is currently an available variant in the protection of metals against corrosion and contributes to the achievement of other special surface properties of materials [11–15]. Together with the applied paints after the galvanizing process, it provides suitable cathodic protection of components exposed to the atmosphere. Identification and analysis of factors acting in this process and monitoring their influence on the formed layer has become the subject of our research and experimental work [4, 10, 16–20] and is also addressed in this monograph. Unlike most scientific and professional works on the issue [13, 14, 21–25], where the selected parameter of the created layer is analysed in relation to only one factor, in our work we focused on a deeper and more comprehensive identification of influence of several factors and their interactions operating in the process of electro-galvanizing based on design of experiments (DOE) methodology. Unlike the COST (consider one separate factor at a time) approach, the DOE methodology allows us to monitor the common influence of several factors on the considered response at a given time and thus find the optimal combination of setting the values of input process variables. In the experimental procedure, the COST approach considers the change of only one separate factor at a time, which is an inefficient approach because it does not provide necessary information to achieve real optimum and thus, experimental work becomes more expensive.

Our goal is to find the optimal duration of the galvanizing process to obtain the desired thickness of the formed protective layer. First, it is necessary to create an optimization model, so we will focus primarily on the development of a mathematical and statistical model predicting the thickness of the layer formed during the process of electroplating with zinc at current density $0.5\left[A \cdot dm^{-2}\right]$. Having applied mathematical and statistical methods and DOE methodology, we will deal with the identification and analysis of factors influencing the galvanizing process. To achieve the set goal, based on DOE methodology with the selected central composite plan, the experiment containing separate experiments was performed and the influence of selected seven input factors on the resulting thickness of the formed zinc layer, which determines the corrosion resistance of the material, was observed. In this part of the monograph, we present the developed mathematical and statistical computational model predicting the thickness of the created layer which was as a basis for optimizing the process by nonlinear programming to save time to increase the efficiency of the whole process with respect to the required layer thickness.

7.1.2 Experimental Part—Galvanizing

The term experiment generally refers to a set of experiments (also measurements and observations) that are arranged in a suitable manner in the case of design of experiment. It can be said that design of experiment creates such conditions that the scope of the experiment is as small as possible, but the volume and information form is of the highest quality and production [26]. The requirement for efficiency is more pronounced in experiments with multiple factors. Its strict respect led to the creation of a separate branch of applied statistics, the so-called design of experiments. Various arrangements of measurements and methods of their evaluation are collectively referred to as design of experiments (DOE) [1–3]. In accordance with the design of experiments methodology, a central composite plan with 40 separate experiments was used to analyse the influence of seven factors on the thickness of the formed zinc layer. Individual levels of independent variables and factors are listed in Table 7.1.

Material S355J0 was used to perform the experiment. The above-mentioned electro-galvanizing at current density $0.5\left[A \cdot dm^{-2}\right]$ was used in individual parts of the experiment. The electrolyte, which is characterized by high depth efficiency, low zinc concentration and high coverage, was used. It is possible to use the same technological equipment as for cyanide baths. Zinc anodes are placed in a separate dissolution bath where it is possible to control zinc content, which is fed to the plating bath through a circulation circuit via a filter using immersion and removal. During alkaline galvanizing, it was necessary to provide the desired concentration of brighteners for the electrolyte as each type of galvanizing bath requires the use of a specific brightener.

Table 7.1 Input technological factors in the galvanizing process

Coded factor	Factor	Unit	Level of factors in the DOE				
			−2.21	−1	0	1	2.21
x_1	*m* (NaOH)	[g·l^{-1}]	19.33	80.00	120.00	180.00	240.67
x_2	*m* (ZnO)	[g·l^{-1}]	3.15	8.00	12.00	16.00	20.85
x_3	*m* (Zn3401)	[ml·l^{-1}]	2.18	4.00	5.50	7.00	8.82
x_4	*m* (Zn3402)	[ml·l^{-1}]	0.68	2.50	4.00	5.50	7.32
x_5	*T*	[°C]	−0.13	12.00	22.00	32.00	44.13
x_6	t	[min]	1.15	6.00	10.00	14.00	18.85
x_7	*U*	[V]	0.79	2.00	3.00	4.00	5.21

Galvanization of the samples was performed based on the DOE methodology [1] with the selected central composite plan with 40 separate realizations of the experiment. We were interested in the influence of seven input factors acting on the thickness of the formed zinc layer, i.e., functional dependence $\hat{y} = f(x_1, x_2, x_3, x_4, x_5, x_6, x_7)$ where

x_1 is the amount of NaOH in the electrolyte,
x_2 is the amount of ZnO in the electrolyte,
x_3 the amount of brightener Pragogal Zn3401 in the electrolyte,
x_4 the amount of brightener Pragogal Zn3402 in the electrolyte,
x_5 electrolyte temperature,
x_6 galvanizing time,
x_7 voltage.

Experimental conditions and individual levels of independent variables (factors) are given in Table 7.1. After the experiment, the thickness of the coated layer was measured on individual samples at preselected experimental points and the measurement was performed using the digital thickness gauge MINITEST 4000. The experimentally obtained data represented the input matrix for further statistical processing.

The DOE methodology used to identify significant factors influencing the thickness of the formed layer, allowed us to obtain the maximum amount of information of high statistical and numerical correctness, thus with high reliability of the realized conclusions [1, 2, 7]. The individual experiments were performed based on the created matrix of DOE as a combination of individual levels of seven input factors in accordance with Table 7.1, where the experimental conditions are presented.

The individual experiments were performed in random order. This randomization is necessary to minimize systematic error or preventing a subjective preference for one of the levels of the input factor. Using scalar products, the orthogonality of the DOE was verified, thus all the columns of the matrix of the DOE must be perpendicular to each other and non-zero to avoid incorrect indication of the statistical insignificance of the regressors [26]. For the DOE standardization (coding) of basic factors, the known conversion relationship [8, 26, 27] was used thanks to which the conversion from the original physical units of factors to a dimensionless form takes place.

7.1.3 Results of Experiment—Mathematical Model Creation

Within statistical analysis of experimentally obtained data, exploratory analysis, screening analysis, analysis of variance, and the DOE analysis were performed using software products, especially MATLAB, Statistica, JMP and QC-Expert. Based on the analysis, we identified significant factors affecting the resulting thickness of the formed zinc layer, analysed their interactions and obtained the form and coefficients of mathematical and statistical models predicting the thickness of the formed layer at varying levels of factors. Data analysis was performed with a statistically correct approach including analysis of basic assumptions and subsequent analysis of the classical regression triplet: data, model, residues. Therefore, we can state that there were no numerical and statistical inaccuracies in the results and conclusions confirmed by practical experience in the field of surface treatment [10, 17, 18, 28–30].

Basic analysis of the achieved results when measuring the thickness of the formed layer in individual experiments results from the analysis of variance (ANOVA) performed in the environment of the Statistica software presented in Table 7.2.

From the ANOVA table (Table 7.2) it can be concluded that the variability caused by random errors is significantly less than the variability of the measured values explained by the model and the value of the achieved significance level (Prob > F) indicates the adequacy of the used model based on the Fisher-Snedecor test criterion. In other words, to test the null statistical hypothesis H_0 which follows from the nature of the test and states that none of the effects used in the model affect a significant change in the examined variable. Consequently, the achieved level of significance ($\alpha = 0.05$) is less than the selected level of statistical significance. Thus, it can be concluded that we do not have enough evidence to accept H_0 and we can say that the model is statistically significant.

Further testing of the used model, the so-called test of the error of insufficient modification of the model, where we monitor the variance of residues and the variance of measured data within groups and thus test whether the regression model sufficiently captures the observed dependence, is shown in Table 7.3. Regarding the achieved significance value of 0.3486 (Prob > F) using the Fisher's test, we can

Table 7.2 ANOVA for the proposed prediction model of the galvanizing process

Source	DF	Sum of squares	Mean square	F ratio	Prob > F
Model	6	207.5383	34.5897	5.9572	0.0003*
Error	33	191.6105	5.8064		
C. total	39	399.1488			

*—significant at level of $\alpha = 0.05$

Table 7.3 Error test of insufficient model fit

Source	DF	Sum of squares	Mean square	F ratio	Prob > F	Max RSq
Lack of fit	28	171.2105	6.11466	1.4987	0.3486	0.9489
Pure error	5	20.4	4.08			
Total error	33	191.6105				

accept the null statistical hypothesis H_0, which results from the nature of the underfit error test, and we can state that the model sufficiently captures the variability of experimentally obtained data. The significance of the model is confirmed in terms of statistical testing, i.e., at the chosen level of significance $\alpha = 0.05$ the variance of residual values is smaller than the variance within individual groups.

Based on the above assumptions and their fulfilment (Tables 7.2 and 7.3), the following table (Table 7.4) presents an estimate of the model parameters with testing the significance of individual effects and their combination at the significance level $\alpha = 0,05$. From the above table of parameter estimates of the used model it is obvious that the thickness of the formed zinc layer is mainly influenced by the applied voltage and the galvanizing time. In addition to the self-acting factors, the square of the amount of sodium hydroxide in the electrolyte and the amount of the gloss forming additive Pragogal Zn3402 also act significantly. However, the *Intercept* acquires the greatest importance, which contains all "neglected" factors in the process of electrolytic alkaline galvanizing. From the point of view of the statistical criterion of nonorthogonality, the *Variance Inflation Factor* (*VIF*) is an important indicator as well as the inflation factors of regressor variance [1, 9, 31].

For *VIF*, the following is valid: platí: less than or equal to one—predictors are not correlated, the plan is uncorrelated and is orthogonal; greater than 1 but less than 5—indication of mild correlation and nonorthogonality of the plan; greater than 5 but less than 10—significant correlation and nonorthogonality of the plan and finally

Table 7.4 Estimation of the model parameters for the galvanizing process

Term	Estimate	Std. error	t Ratio	Prob > |t|	Lower 95%	Upper 95%	*VIF*
Interc	14.10792	0.599205	23.5	<0.0001*	12.88882	15.32701	
x_7	1.483657	0.432078	3.43	0.0016*	0.604588	2.362725	0.985741
x_6	1.043233	0.432078	2.41	0.0215*	0.164164	1.922301	0.985741
$x_1 \cdot x_1$	−1.06999	0.368621	−2.9	0.0065*	−1.81995	−0.32002	0.911005
$x_1 \cdot x_2$	−0.78191	0.512519	−1.53	0.1366	−1.82464	0.260817	0.985741
$x_5 \cdot x_4$	0.647948	0.512519	1.26	0.215	−0.39478	1.690676	0.985741
$x_4 \cdot x_4$	−0.79972	0.368621	−2.17	0.0373*	−1.54969	−0.04976	0.911005

x_7—voltage in volts, x_6—galvanizing time in minutes, x_1—amount of NaOH in [g·l^{-1}]
x_2—amount of ZnO in [g·l^{-1}], x_4—amount of brightener Pragogal Zn 3402 v [ml·l^{-1}]
x_5—electrolyte temperature in [°C], intercept—absolútny member of the model, *—factor or interaction of factors is significant at the level of significance 5%

greater than 10—multicollinearity of regressors and nonorthogonality of the plan. In case where *VIF* > 1 the regression coefficient estimates are numerically correct but their *p—values,* which are determined as diagonal elements of the inverse correlation matrix ($j = 1, \ldots, p$), are not correct.

$$VIF_j = D_{jj} = diag(\mathbf{D}) = diag(\mathbf{R}^{-1}) \tag{7.1}$$

Given that the results indicate the non-correlation of predictors and the orthogonality of the experimental design and based on Table 7.4 we can compile a prediction equation for the thickness of the formed layer $\hat{y} = th$ in a coded form:

$$\begin{aligned} \hat{y} &= 14.10792 + 1.483657\, x_7 + 1.043233\, x_6 - 1.06999\, x_1^2 \\ &\quad - 0.78191\, x_1\, x_2 + 0,647948\, x_5\, x_4 - 0.79972\, x_4^2 \end{aligned} \tag{7.2}$$

To compile the prediction relationship in nature unit, it is necessary to realize that during the analysis, factors in coded unit were used, coded by the DOE standardization according to the relationship

$$x_d(i) = \frac{x(i) - \frac{x_{\max}+x_{\min}}{2}}{\frac{x_{\max}-x_{\min}}{2}} \tag{7.3}$$

where $x_d(i)$—is the standardized variable according to the DOE, $x(i)$—is the original basic variable, where $i = 1, 2, \ldots, n$, where n is the number of basic factors, $x_{\max}$—is the maximum value of the original variable $x(i)$, $x_{\min}$—is the minimum value of the original variable $x(i)$. The DOE standardization according to (7.3) is a linear transformation of factor values from the original physical units to a dimensionless form, i.e., from the interval $\langle x_{\min}, x_{\max}\rangle$ to the interval $\langle -1, 1\rangle$. Considering the conversion relation (7.3) and the statistical prediction Eq. (7.2), it is possible to write a prediction relation describing the investigated dependence for the current density $0.5\left[\mathrm{A}\cdot\mathrm{dm}^{-2}\right]$ in natural unit in the following form

$$\begin{aligned} th &= m(Zn3402)\cdot 0.386 - 0.035\, T + 0.671\, U + 0.118\, t \\ &\quad + 0.034\, m(NaOH) + 0.112\, m(ZnO) - 8.736\cdot 10^{-5}\,(m(NaOH))^2 \\ &\quad - 0.073\,(m(Zn3402))^2 - 7.983\cdot 10^{-4}\, m(NaOH)\cdot m(ZnO) \\ &\quad + 8,819\cdot 10^{-3}\, m(Zn3402)\cdot T + 6.806 \end{aligned} \tag{7.4}$$

The prediction Eq. (7.3) will serve as a basis for optimizing the galvanizing process using nonlinear programming.

7.1.4 Optimization of the Galvanizing Process in MATLAB

When optimizing surface treatment processes, the essence of optimization problems lies in determining such a combination of values of individual control factors entering the galvanizing process in which we get the best value of the optimization parameter expressed through the optimality criterion included in the objective function. The obtained values of factors after the optimization procedure are called the optimal values. In terms of metals surface treatment, one of the most important variables determining the efficiency of the whole process is time, i.e., the duration of the process itself. If we can minimize the time required to create a layer with the required thickness when setting the factors, we can maximize the economic yield while ensuring the required quality. From the Eq. (7.4) we express time t, i.e., galvanizing time to obtain the objective, i.e., criterion function:

$$\begin{aligned} t = {} & 8.475\,th - 3.271\,m(Zn3402) + 0.297\,T - 5.686\,U \\ & - 0.288\,m(NaOH) - 0.949\,m(ZnO) + 7.403 \cdot 10^{-4}\,(m(NaOH))^2 \\ & + 0.619\,(m(Zn3402))^2 + 6.765 \cdot 10^{-3}\,m(NaOH) \cdot m(ZnO) \\ & - 7.474 \cdot 10^{-2}\,m(Zn3402) \cdot T - 57.678 \end{aligned} \tag{7.5}$$

The basic goal of nonlinear optimization in MATLAB environment is to find the minimum (local extremum) of the objective function for the optimization problem written in a format suitable for the given software, i.e., in the following form:

$$\min_x f(x) \begin{cases} c(x) \le 0 \\ ceq(x) = 0 \\ \mathbf{A} \cdot \mathbf{x} \le \mathbf{b} \\ \mathbf{Aeq} \cdot \mathbf{x} = \mathbf{beq} \\ \mathbf{lb} \le x \le \mathbf{ub} \end{cases} \tag{7.6}$$

where **x**, **b**, **beq**, **lb** (*lower boundary*) and **ub** (*upper boundary*) are vectors, **A** and **Aeq** are matrices with constant coefficients, $c(x)$ and $ceq(x)$ are vector functions and $f(x)$ is a scalar function.

Functions $f(x), c(x)$ and $ceq(x)$ are nonlinear [32]. When defining the constraints of the electrolytic alkaline galvanizing process considering the data given in Tables 7.1 and 7.4 and the nature of the process, we get constraints in the form of inequalities:

$$80 \le m(NaOH) \le 130 \tag{7.7}$$

$$7.5 \le m(ZnO) \le 18 \tag{7.8}$$

$$2 \leq m(\text{Pragogal } Zn \ 3402) \leq 6.5 \tag{7.9}$$

$$12 \leq T \leq 28 \tag{7.10}$$

$$2 \leq U \leq 5 \tag{7.11}$$

To run the optimization procedure for solving the optimization problem with the objective function (7.5) and constraints (7.7)–(7.11), we created the appropriate m.file in MATLAB using nonlinear programming. Considering the required layer thickness 12 [μm], which is the most often required thickness during galvanizing as well as considering technological constraints (7.7)–(7.11), we obtained the optimal time $t = 11.305$ [min] for galvanizing after the implementation of the optimization procedure. It is possible with the optimal combination of individual control variables, i.e., when setting the input factors at optimal values:

$$m(NaOH) = 112.585 \left[\text{g} \cdot \text{l}^{-1}\right], m(ZnO) = 18 \left[\text{g} \cdot \text{l}^{-1}\right], T = 12 \left[{}^{\circ}\text{C}\right],$$

$$m(Pragogal\ Zn\,3402) = 3.392 \left[\text{ml} \cdot \text{l}^{-1}\right] \text{and } U = 5 \ [\text{V}].$$

Graphical output of the optimization is shown in Figs. 7.1, 7.2, 7.3, 7.4 and 7.5.

In MATLAB environment, we can verify whether any of the technological constraints has not been violated. This is presented in the following graphic output of the optimization procedure in Fig. 7.3.

When analysing the galvanizing technological process, when it is important to simultaneously monitor the interaction of several factors, it is more appropriate to apply the DOE approach in experimentation, which we present in this chapter to solve a specific problem in practice. Using the DOE and based on statistical analysis of experimentally obtained data in the process of electrolytic alkaline galvanizing, we compiled a predictive mathematical and statistical model predicting the thickness

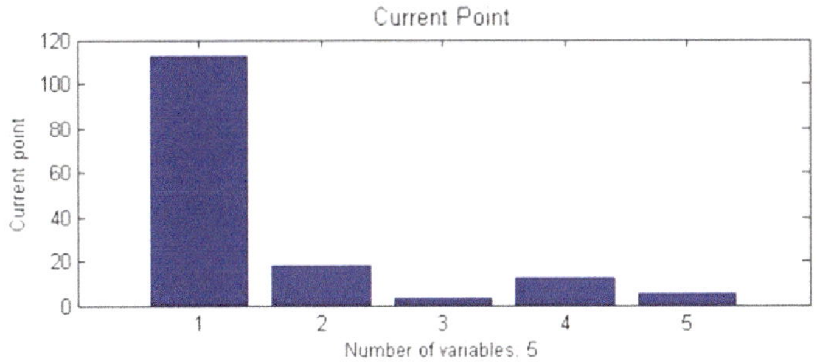

Fig. 7.1 Graph of existing galvanizing time optimization variables for the required layer thickness 12 [μm]

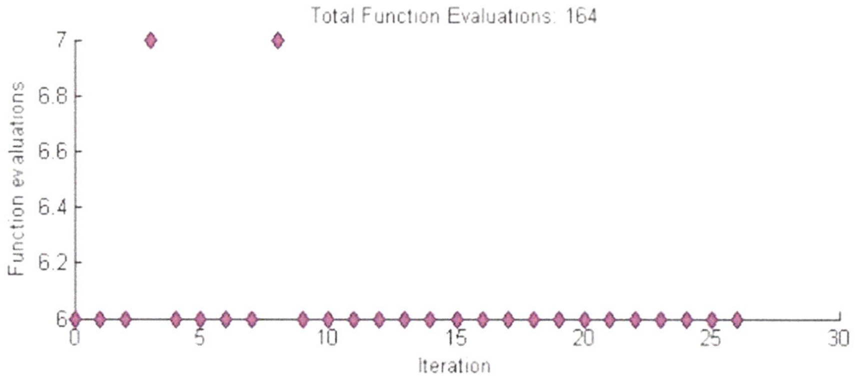

Fig. 7.2 Graphic output of the galvanizing time optimization procedure for the required layer thickness 12 [μm]

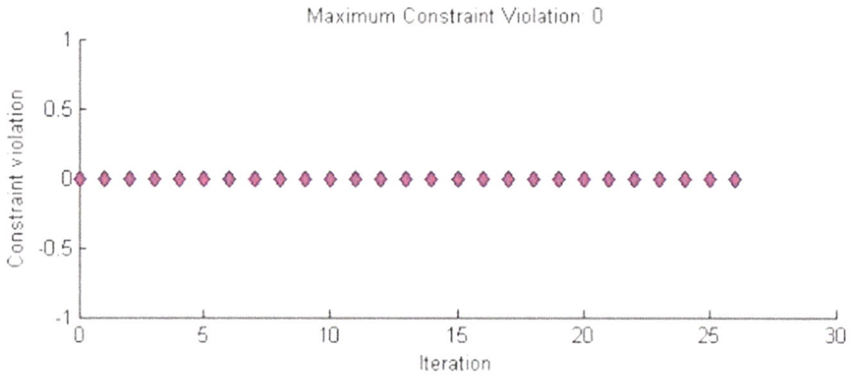

Fig. 7.3 Graphic output of the galvanizing time optimization procedure for the required layer thickness 12 [μm]—maximum number of constraint violations

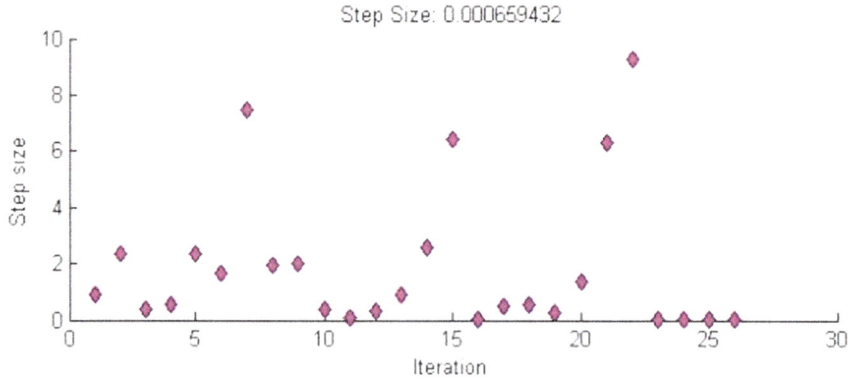

Fig. 7.4 Graphic output of the galvanizing time optimization procedure for layer thickness 12 [μm]—step size in iterations during optimization

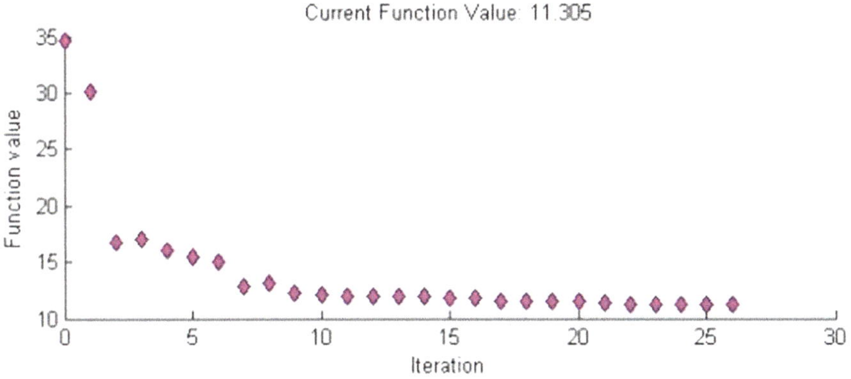

Fig. 7.5 Graphic output of the galvanizing time optimization procedure for the layer thickness of 12 [μm], the achieved optimal galvanizing time t = 11.305 [min]

of the resulting protective layer. Subsequently, by applying nonlinear programming, we optimized the duration of the galvanizing process. The main benefit for engineering practice is that the optimal process conditions of the given surface treatment were calculated, thanks to which it was possible to ensure compliance with the requirements of end customers while increasing the efficiency of the process from an economic point of view.

7.2 Application of Nonlinear Programming to Optimize the Process of Anodic Aluminium Oxidation

In this chapter we will focus on application of nonlinear programming to optimize the technological process of *anodic aluminium oxidation* (AAO), the so-called anodizing. In engineering practice, it often happens that industrial or technological processes are not implemented under optimal conditions. The process is usually influenced by many factors that interact and thus affect the course of the process in a non-additive way. In this case, the process is a complex, multifactorial, and nonlinear system. In technological and technical practice, it is an important and frequent goal to find out the connections and relationships between certain quantities of the investigated process. For processes that are very complex and there is no (yet) description or suitable mathematical, physical, and chemical model for them. It is necessary to recognize and identify these relationships experimentally. The main and most common goal of the implemented technical and technological experiments is to determine whether certain factors (input quantities) have an impact on the monitored output quantity (often called response). This is followed by the goal of finding such levels of factors that the optimum (maximum, minimum) of the observed quantity is achieved [1–3, 8, 33, 34, 35]. This was stated for the galvanizing process and the same is the case with another of the surface treatment processes—the process of *anodic aluminium oxidation* (AAO).

AAO is one of the most efficient and most used surface treatment processes, because with the right choice (setting) of technological factors it is possible to create a protective layer of material (components) with the required thickness and properties that meets all defined criteria (e.g., hardness, resistance to wear, corrosion resistance, etc.). A persistent problem in AAO is that despite many academic and industrial studies [35–41], a unified theory and a clear mechanism of layer formation and its growth have not been established yet. Therefore, the identification and analysis of the factors involved in this process and monitoring their impact on the created layer to optimize the operation of the process is a very important goal in scientific research. It has been a long-term subject of our scientific research and experimental work [20, 42–49].

Unlike most scientific and professional works in the field, where the selected layer parameter is analysed in relation to only one factor [50–55], we focused on a deeper and more comprehensive identification of the influence of several factors and their interactions in the process of AAO, which did not go without the use of the methodology of the DOE. It is analogous to the galvanizing process. It is obvious that neglecting certain principles of the DOE could lead to the devaluation of the whole experimental work (within experiments in the field of surface treatment it is time consuming and economically demanding), the DOE is an ideal choice because even with a lower number of experiments their information form is of high possible quality [1, 6, 27, 49, 56, 57]. Based on the data analysis obtained experimentally in AAO in sulphuric acid at constant current density 1 $\left[\mathrm{A} \cdot \mathrm{dm}^{-2}\right]$ and 2 $\left[\mathrm{A} \cdot \mathrm{dm}^{-2}\right]$ direct current (GS method), our goal was to identify significant factors affecting the thickness of the formed AAO layer and investigate the influence of main effects and their interactions. Furthermore, it was important to create mathematical and statistical models based on which it is possible to predict the thickness of the created anodic layer (basic customer requirement) for the considered factors and their interactions. The developed models will again form the basis for optimizing the AAO process and finally for determining the critical temperature.

7.2.1 Experimental Part—Anodic Aluminium Oxidation

During the experiment, aluminium samples EN AW-1050A H24 with the dimensions $100 \times 70 \times 0.5$ [mm] were used as experimental material. The above-mentioned GS method of AAO (in sulphuric acid at direct current) at the selected constant current density was used in individual parts of the experiment. The individual experiments were performed in a Hull bath, which has the characteristic shape of a trapezoid [48, 58]. The connected experimental apparatus was equipped with an adjustable direct current source.

Anodic oxidation of aluminium samples itself took place based on the DOE methodology with the selected central composite design with 26 separate experimental implementations. Due to the assumed nonlinear dependencies, a central composite design (CCP) was chosen from a relatively large number of types of designes used in the DOE. Individual experiments were performed based on the created matrix of the experimental design as a combination of individual levels of four input factors in accordance with Table 7.5, where the experimental conditions are presented. The randomization of the order of the individual experiments and the orthogonality of the experimental design were also observed in the case of the AAO process. The coding of DOE factors by standardization to the dimensionless form according to the relation (7.3) was also ensured.

The influence of input chemical factors x_1, x_2 and physical factors x_3, x_4 and their interactions on the resulting thickness $\hat{y}$ of the formed oxide layer at a constant anodic current density of 1 $[A \cdot dm^{-2}]$ and 2 $[A \cdot dm^{-2}]$ was investigated. Functional dependency $\widehat{y} = f(x_1, x_2, x_3, x_4)$ was sought, where

x_1 the amount of sulphuric acid H_2SO_4 in the electrolyte,
x_2 concentration of aluminium cations Al in the electrolyte,
x_3 electrolyte temperature T,
x_4 time t expressing the duration of AAO process (plating time).

Functional dependence $th = f(m(H_2SO_4), m(Al), T, t)$ is sought in nature unit of the individual factors. Experimental conditions and individual levels of independent variables (factors) are given in Table 7.5.

After the experiment, the thickness of the coated layer was measured on individual samples at the preselected experimental points. Measurement was performed using a

Table 7.5 Experimental conditions in the AAO process

Experiment GS							
AAO method				GS			
Type of DOE design				Central composite design (CCD)			
Number of factor used			4	**Experimental trials**			26
Controlled factors							
Coded factor	**Factor in nature unit**	**Unit**	**Level of factors**				
			−2	−1	0	1	2
x_1	m (H_2SO_4)	$[g \cdot l^{-1}]$	25.00	100.00	175.00	250.00	325.00
x_2	m (Al)	$[g \cdot l^{-1}]$	1.32	3.97	6.62	9.26	11.91
x_3	T	[°C]	2.00	12.00	22.00	32.00	42.00
x_4	t	[min]	10.00	25.00	42.50	60.00	80.00
Constant factors							
Anode/Cathode material		AW – 1050A H24		**Voltage**		12 V	

digital thickness gauge MINITEST 4000. Experimental points were at the intersections of horizontal and vertical lines, the distance between the lines from the bottom edge of the sample was 5 [mm]. The thickness of the formed layer was measured at each point 5 times. To prevent a systematic measurement error, the measurement was also performed on the standard. The arithmetic mean of the five measurements, which was calculated by the thickness gauge microprocessor, was taken as an individual measurement. The experimentally obtained data represented the input matrix for further statistical processing to develop a predictive mathematical-statistical model.

7.2.2 Results of the Experiment—Mathematical Modelling

Based on statistical analysis of experimentally obtained data (exploratory data analysis EDA, screening analysis, ANOVA, and the DOE analysis) using software products (MATLAB, Statistica, JMP, QC-Expert), we have identified significant factors affecting the resulting thickness of the AAO layer. We analysed the interactions of the investigated factors and obtained the resulting shape and coefficients of mathematical and statistical models predicting the thickness of the formed layer at changing levels of factors. Data analysis was performed with a statistically correct approach including analysis of basic assumptions and subsequent analysis of the classical regression triplet: data, model, residuals. Therefore, we can state that there were no numerical and statistical inaccuracies in the results and conclusions, which was confirmed by practical experience in the field of surface treatment [20, 45, 47, 48, 59–63].

The screening analysis showed that at the chosen level of significance $\alpha = 0.05$, amount of aluminium in the electrolyte as a single acting factor has no significant effect on the thickness of the layer formed at the anodic current density $J_A = 1\ [\mathrm{A} \cdot \mathrm{dm}^{-2}]$ and $J_A = 2\ [\mathrm{A} \cdot \mathrm{dm}^{-2}]$ only in interaction. Predictive models with the DOE transformed variables were developed for the investigated factors x_1, x_2, x_3, x_4 and their interactions. However, we know from practice that the factor x_2 influences the deterioration of the quality of the formed layers. Statistical analysis of the suitability of individual models of the factors effect on the thickness of the formed layer using GS method showed that the share of variability of the measured thicknesses expressed by *RSquare* (coefficient of determination R^2) represents a value of 93.17% for current density $J_A = 1\ [\mathrm{A} \cdot \mathrm{dm}^{-2}]$ and 91.52% for $J_A = 2\ [\mathrm{A} \cdot \mathrm{dm}^{-2}]$. Adjusted coefficient of determination denoted as *Rsquare Adj* (or $Adj\mathrm{R}^2$) expressing the degree of suitability of the use of the given model for the description of the relationship between controlled (chemical and physical) factors and the observed investigated system response, i.e., thickness of the formed layer, represents a value of $Adj\mathrm{R}^2 = 90.52\%$ for the current density $J_A = 1\ [\mathrm{A} \cdot \mathrm{dm}^{-2}]$. This points to the fact that up to 90.52% of the variability of the data (experimentally obtained thicknesses of the formed layer) is explained by the prediction model, i.e., the dependence between the observed variables can be considered functional according to the Cohen scale. For $J_A = 2\ [\mathrm{A} \cdot \mathrm{dm}^{-2}]$ *Rsquare Adj* acquires a value of 88.22%.

From the ANOVA tables for individual current densities (Tables 7.6, 7.7) it can be concluded that variability caused by random errors is significantly less than variability of measured values explained by the model and the value of the achieved significance level ($Prob > F$) points to the adequacy of the used model based on the Fisher-Snedecor test criterion.

To test the null statistical hypothesis H_0, which is based on the nature of the test and states that none of the effects used in the model affect a significant change in the examined variable, it follows that the achieved level of significance ($Prob > F$) is less than the selected level of statistical significance $\alpha = 0.05$. Thus, it can be concluded that we do not have enough evidence to accept H_0 and we can say that the model is statistically significant. This is confirmed by the results of *ANOVA* for individual local current densities, on the basis of which it can be concluded that at the selected level of significance $\alpha = 0.05$ there is at least one factor that significantly affects the resulting layer thickness formed by AAO at local anode current density and thus based on the achieved level of significance the value of which is less than 0.0001, the chosen model can be considered adequate.

Testing of the insufficient adjustment error of the model, in which the variance of residues and the variance of the measured data within the groups are evaluated, thus testing whether the regression model sufficiently captures the observed dependence, is given in (Table 7.8).

Regarding the achieved value of significance using the 0.1973 Fisher test, we can accept a null statistical hypothesis at the chosen significance level $\alpha = 0.05$ for current density $J_A = 1\ \left[A \cdot dm^{-2}\right]$ which results from the nature of the lack-of-fit error test. We can state that the model sufficiently captures the variability of experimentally obtained data because the variance of residual values is smaller than the variance

Table 7.6 ANOVA for AAO process prediction model at $J_A = 1[A{\cdot}dm^{-2}]$

Source	DF	Sum of squares	Mean square	F ratio	Prob > F
$J_A = 1\ [A{\cdot}dm^{-2}]$					
Model	7	636.44441	90.9206	35.0963	<0.0001*
Error	18	46.63094	2.5906		
C. total	25	683.07536			

*—significant at level of $\alpha = 0.05$

Table 7.7 ANOVA for AAO process prediction model at $J_A = 2[A{\cdot}dm^{-2}]$

Source	DF	Sum of squares	Mean square	F ratio	Prob > F
$J_A = 2\ [A{\cdot}dm^{-2}]$					
Model	7	703.89575	100.557	27.7562	<0.0001*
Error	18	65.2113	3.623		
C. total	25	769.10704			

*—significant at level of $\alpha = 0.05$

Table 7.8 Test of errors of lack-of-fit of AAO models

Source	DF	Sum of squares	Mean Square	F ratio	Prob > F	Max RSq
$J_A = 1$ [A·dm^{-2}]						
Lack of fit	7	24.512724	3.50182	1.7416	0.1973	0.9676
Pure error	11	22.118219	2.01075			
Total error	18	46.630944				
$J_A = 2$ [A·dm^{-2}]						
Lack of fit	7	41.674835	5.95355	2.7824	0.0629	0.9694
Pure error	11	23.536462	2.13968			
Total error	18	65.211297				

within individual groups. The same conclusions based on the results given in Table 7.8 can also be accepted for current density $J_A = 2\left[A \cdot dm^{-2}\right]$.

It is obvious from the table of the model Estimate (Table 7.9) that the resulting value of the thickness of the formed AAO layer at $J_A = 1\left[A \cdot dm^{-2}\right]$ is influenced by the amount of sulphuric acid in the electrolyte as a representative of chemical factors and both physical factors (electrolyte temperature and anodic oxidation time). A significant effect of electrolyte concentration, exposure time and temperature can be observed here. Factor interactions and the *intercept*, which contains all the "neglected" influences of the process, are also important. The most significant effect (if we do not consider the intercept) has the amount of sulphuric acid, the separate effect of which is given by the positive sign of the estimate of the individual t-test. However, the overall effect must also be examined in terms of significant effect interactions, where increasing the total amount of acid in conjunction with the time of anodic oxidation acts in the opposite direction, thus leading to a reduction in the thickness of the formed layer by dissolving it.

Table 7.9 Estimation of model parameters of anodizing at $J_A = 1$[A·dm^{-2}]

Term	Estimate	Std. error	t ratio	Prob > |t|	Lower 95%	Upper 95%	VIF
Interc	12.193	0.430	28.35	<0.0001*	11.290	13.097	–
x_1	0.385	0.569	0.68	0.5069	−0.810	1.581	3
x_4	4.047	0.569	7.11	<0.0001*	2.852	5.243	3
x_3	1.817	0.329	5.53	<0.0001*	1.127	2.507	1
$x_3 \cdot x_3$	−1.972	0.317	−6.23	<0.0001*	−2.637	−1.307	1
$x_1 \cdot x_1 \cdot x_1$	1.163	0.232	5	<0.0001*	0.674	1.651	3
$x_1 \cdot x_1 \cdot x_4$	−2.089	0.697	−3	0.0077*	−3.553	−0.625	3
$x_1 \cdot x_2 \cdot x_3$	−1.203	0.402	−2.99	0.0079*	−2.048	−0.357	1

x_1—amount of H_2SO_4 (in natural unit in [g·l^{-1}]), x_2—amount of Al in the electrolyte (natural unit [g·l^{-1}]), x_3—electrolyte temperature (natural unit [°C]), x_4—AAO deposition time (in [min]), *—significant at level of $\alpha = 0.05$

Based on the estimates of the model parameters (Table 7.9), it is possible to compile a mathematical and statistical model of the prediction dependence for DOE standardized variables in the following form:

$$\hat{y}_{J_A = 1\,[\mathrm{A \cdot dm^{-2}}]} = 12.19 + 4.05\,x_4 + 1.82\,x_3 - 1.97\,x_3^2 + x_1^2(1.16\,x_1 - 2.09\,x_4) - 1.20\,x_1\,x_2\,x_3 \quad (7.12)$$

Using the conversion relation (7.3) and the model Eq. (7.12), it is possible to transform the given statistical model into nature units of individual factors (independent variables) and thus create a technological calculation model predicting the resulting thickness th of the created layer:

$$\begin{aligned} th_{J_A=1\,[\mathrm{A \cdot dm^{-2}}]} &= 5.93 + 0.22\,T - 9.69 \cdot 10^{-3}\,t - 0.02\,m(H_2SO_4) \\ &- 6.8 \cdot 10^{-5}[m(H_2SO_4)]^2 + 3.44 \cdot 10^{-7}[m(H_2SO_4)]^3 + 2 \cdot 10^{-3}\,T\,t \\ &+ 4.87 \cdot 10^{-4} m(H_2SO_4)\,T + 1.18.10^{-3} m(H_2SO_4)\,t - 4.93 \cdot 10^{-3}\,T^2 \\ &- 2.65 \cdot 10^{-6}[m(H_2SO_4)]^2\,t - 1.15 \cdot 10^{-5} m(H_2SO_4)\,T\,t \end{aligned} \quad (7.13)$$

The individual factors used to create the mathematical and statistical model are presented in Table 7.9 sorted according to the significance of the influence of the given factor itself or the interaction of several factors (size of the absolute value of test statistics *t Ratio*—Student's test statistics). The absolute term of model denoted as *Intercept* (Table 7.9) includes all unconsidered influences, i.e., the type of the experiment design, the use of independent variables themselves such as neglect of other influences (material impact) and the levels of adjustment used for each factor. Due to the complexity of the performed analysis and confirmation of the correctness and suitability of the chosen model, it is required to verify the residues, i.e., the difference between the measured and predicted values (calculated using the prediction model) in terms of their distribution and autocorrelation. The value of the Durbin-Watson autocorrelation test $D_W = 2.0378084$ allows to accept a zero statistical hypothesis of the absence of autocorrelation. The achieved level of significance of the Shapiro-Wilks test (Fig. 7.6) indicates Gaussian distributions of residues. In conclusion, we can evaluate the prediction model as statistically and numerically correct.

Analogously, a mathematical-statistical calculation model was created for the current density $J_A = 2\left[\mathrm{A \cdot dm^{-2}}\right]$ based on which it is possible to predict the resulting thickness of the formed AAO layer during anodic oxidation using GS method. The individual factors used to create the mathematical and statistical model for current density $J_A = 2\left[\mathrm{A \cdot dm^{-2}}\right]$ are listed in Table 7.10, where they are reordered according to the significance of the impact. The most important member is the intersect, which includes neglected and non-considered influences. As in the previous model, at this current density it can be observed that the amount of dissolved aluminium in the electrolyte as a factor acting individually does not have a significant effect on the thickness of the formed layer, but it is still necessary to consider its interaction and limit value 12 $\left[g \cdot \mathrm{l}^{-1}\right]$ observed in terms of real production. When

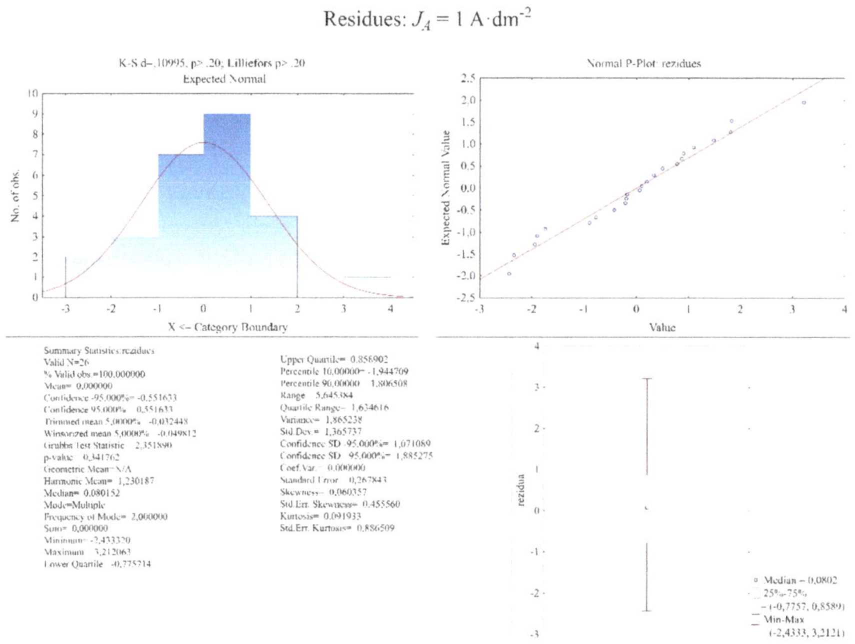

Fig. 7.6 Residue analysis of prediction model for $J_A = 1\ [\text{A} \cdot \text{dm}^{-2}]$

this value is exceeded, the quality of the formed layer begins to deteriorate significantly and, on the contrary, it dissolves. Based on the analysis, we can state the most significant effect of the duration of anodic oxidation for the local anode current density $J_A = 2\ [\text{A} \cdot \text{dm}^{-2}]$ presented by a factor x_4 that also occurs in significant interactions together with the amount of sulphuric acid in the electrolyte.

Table 7.10 Estimation of model parameters of anodizing at $J_A = 2[\text{A}{\cdot}\text{dm}^{-2}]$

Term	Estimate	Std error	t ratio	Prob > |t|	Lower 95%	Upper 95%	VIF
Interc.	13.560675	0.508699	26.66	<0.0001*	12.491937	14.629412	
x_4	2.9105093	0.388526	7.49	<0.0001*	2.0942474	3.7267711	1
x_1	4.9372222	0.672946	7.34	<0.0001*	3.5234153	6.3510292	3
x_3	1.9113426	0.388526	4.92	0.0001*	1.0950808	2.7276044	1
$x_1 \cdot x_3$	−1.316875	0.475845	−2.77	0.0127*	−2.316587	−0.317163	1
$x_3 \cdot x_3$	−2.189342	0.374393	−5.85	<0.0001*	−2.975912	−1.402772	1
$x_4 \cdot x_4 \cdot x_1$	−3.573125	0.824187	−4.34	0.0004*	−5.304678	−1.841572	3
$x_1 \cdot x_3 \cdot x_4$	−1.510347	0.475845	−3.17	0.0053*	−2.51006	−0.510635	1

x_1—amount of H_2SO_4 (in natural unit in [g·l^{-1}]), x_2—amount of Al in the electrolyte (natural unit [g·l^{-1}]), x_3—electrolyte temperature (natural unit [°C]), x_4—AAO deposition time (in [min]), *—significant at level of $\alpha = 0.05$

Based on the estimates of the model parameters (Table 7.10), it is possible to compile for $J_A = 2\ [\mathrm{A \cdot dm^{-2}}]$ the predictive mathematical and statistical model the dependence of the coated layer thickness on the input factors, i.e., for DOE standardized variables model thickness in the form

$$\hat{y}_{J_A = 2\ [\mathrm{A \cdot dm^{-2}}]} = 13.56 + 2.91\, x_4 + 4.94\, x_1 + 1.91\, x_3 - 1.32\, x_1\, x_3 - 2.19\, x_3^2 - 3.57\, x_4^2\, x_1 - 1.51\, x_1\, x_3\, x_4 \tag{7.14}$$

Using the conversion relation (7.3) and the model Eq. (7.14) it is possible to transform the given statistical model into nature units of individual factors (independent variables) and then create a technological calculation model predicting the resulting thickness th of the created layer:

$$\begin{aligned} th_{J_A=2\ [\mathrm{A \cdot dm^{-2}}]} &= 6.33 + 0.31\, T - 6 \cdot 10^{-3}\, m(H_2SO_4) - 0.26\, t \\ &+ 1.97 \cdot 10^{-3}\, m(H_2SO_4)\, t + 2.52 \cdot 10^{-3}\, T\, t \\ &- 1.94 \cdot 10^{-5}\, m(H_2SO_4)\, t^2 - 5.47 \cdot 10^{-3}\, T^2 + 3.4 \cdot 10^{-3}\, t^2 \\ &+ 1.72 \cdot 10^{-4}\, m(H_2SO_4)\, T - 1.44 \cdot 10^{-5}\, m(H_2SO_4)\, T\, t \end{aligned} \tag{7.15}$$

Due to the complexity of the assessment of the correctness and suitability of the technological prediction model (7.15), a residue analysis was performed. The Durbin-Watson autocorrelation test with $D_W = 2.0378084$ indicates that there is no autocorrelation (Fig. 7.7).

7.2.3 *Optimization of Anodic Aluminium Oxidation Process in MATLAB*

To optimize anodic aluminium oxidation (AAO), we have created two models for the most used anode current densities $J_A = 1\ [\mathrm{A \cdot dm^{-2}}]$ and $J_A = 2\ [\mathrm{A \cdot dm^{-2}}]$ based on the performed experiment and statistical analysis. These two models are defined by the mathematical and statistical relations (7.12) and (7.14) for variables in coded unit (DOE coding) and the relations (7.13) and (7.15) for variables in nature unit.

Deposition time was selected to be an optimality criterion, i.e., AAO layer formation time or process duration. The optimality criterion considers the aspect of economic efficiency of the given surface treatment process because in AAO the decisive indicator of economic efficiency is deposition time, which is the longest in comparison with the times of other technological operations (degreasing, pickling, clearing, sealing), and this one defines the cycle of the production line. If we minimize the time required to form a layer with the required thickness by setting the optimal combination of values of the acting factors, we can maximize the economic yield while ensuring the required quality.

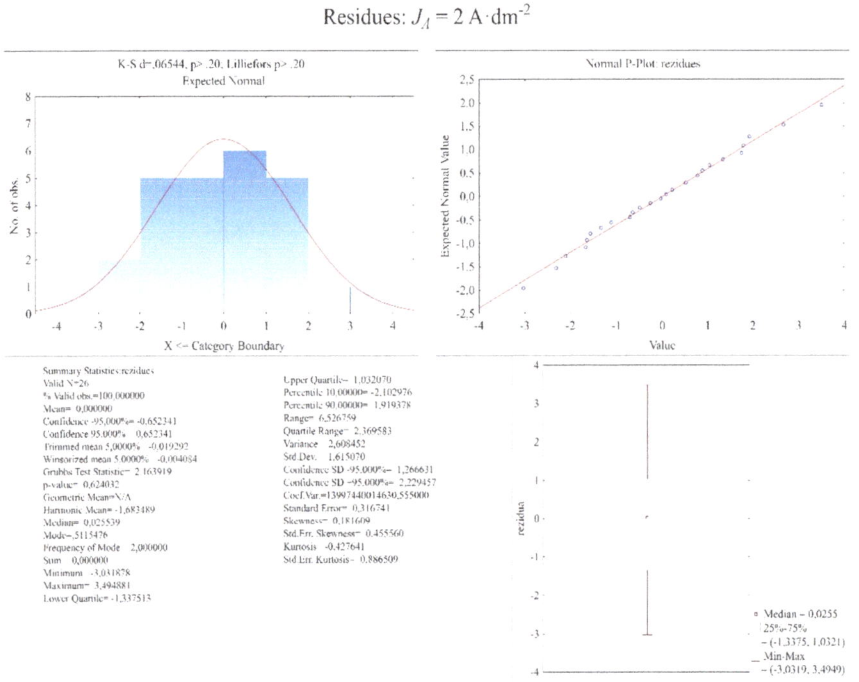

Fig. 7.7 Residue analysis of prediction model for $J_A = 2\ \left[\mathrm{A} \cdot \mathrm{dm}^{-2}\right]$

From practical experience, it can be said that up to 95% of customers define the thickness of the created layer as a basic requirement in AAO surface treatment. It is the most entered parameter in a design documentation. Therefore, as part of the optimization procedure, our objective will be to minimize the deposition time for a predetermined thickness of the formed final layer.

In the theoretical part of the monograph in Sect. 2.2, we have defined the problem of mathematical programming in the narrower sense by the relation (2.4) and in the broader sense by the relation (2.6). In Sect. 6.3, which is devoted to the optimal choice of cutting parameters, we have mentioned it once again as a minimization problem of mathematical programming in the broader sense:

$$Min\left\{f_0(x)\ |x \in X, f_i(x) \leq 0, i \in I,\ h_j(x) = 0,\ j \in J\right\} \tag{7.2.6}$$

with constraints in the form of inequalities and equations, where I, J are index sets. If at least one of the functions f_0, f_i, $i \in I$, h_j, $j \in J$ is not linear, we speak of an optimization problem called ***nonlinear programming***. We will apply nonlinear programming in case of optimizing the AAO process as the objective function (expressing the functional dependence for the deposition time) will be nonlinear.

In our case, the basic goal of optimization can be defined as minimization of deposition time t while maintaining the prescribed thickness th_p of the resulting layer which can be written in the general form:

$$t\,(x_4) = f_0(x_1,\, x_2,\, x_3)_{th=th_p} \quad \rightarrow \quad \min \tag{7.16}$$

where

x_1 expresses the amount of H_2SO_4 in the electrolyte,
x_2 expresses the amount of Al in the electrolyte,
x_3 expresses the temperature of the electrolyte,
$t\,(x_4)$ AAO layer deposition time in nature unit in minutes.
th hrúbka vytvorenej AAO vrstvy ako výsledok procesu.
th_p predpísaná hrúbka vytváranej AAO vrstvy.

The Eq. (7.16) expresses the objective function $f_0(\mathbf{x})$ that models time saving. Our goal is to find such a combination x_1, x_2, x_3 from feasible region $(x_1, x_2, x_3) = \mathbf{x}^* \in X$ which will guarantee the minimum value of the objective (criterion) function $f_0(\mathbf{x})$ while maintaining all the prescribed technological constraints. We will call this combination of numbers the optimal solution.

To simplify calculations, we will follow from the mathematical and statistical model (7.12) which defines the thickness of the created layer at the current density $J_A = 1\ \left[\mathrm{A} \cdot \mathrm{dm}^{-2}\right]$ in a coded form:

$$\hat{y}(th)_{J_A=1\,[\mathrm{A}\cdot\mathrm{dm}^{-2}]} = 12.19 + 4.05\,x_4 + 1.82\,x_3 - 1.97\,x_3^2 + x_1^2(1.16\,x_1 - 2.09\,x_4) - 1.20\,x_1\,x_2\,x_3 \tag{7.12}$$

In the following, we will denote the left side of the dependency (7.12) simply as th.

$$th = 12.19 + 4.05\,x_4 + 1.82\,x_3 - 1.97\,x_3^2 + 1.16\,x_1^3 - 2.09\,x_1^2\,x_4 - 1.20\,x_1\,x_2\,x_3 \tag{7.12*}$$

By simple adjustments from the dependence (7.12*) we will express the factor x_4, in which the deposition time t of the AAO layer is included, i.e.

$$x_4 = \frac{th - 1.82\,x_3 + 1.97\,x_3^2 - 1.16\,x_1^3 + 1.20\,x_1\,x_2\,x_3 - 12.19}{4.05 - 2.09\,x_1^2} \tag{7.17}$$

For transformation to nature unit, we use the relation (7.3) and considering the experimental intervals for individual factors, which are listed in Table 7.5 (experimental conditions), we get the transformation relations for individual variables:

$$x_1 = \frac{m(H_2SO_4) - 175}{150} \tag{7.18}$$

$$x_2 = \frac{m(Al) - 6.62}{5.295} \tag{7.19}$$

$$x_3 = \frac{T - 22}{20} \tag{7.20}$$

$$x_4 = \frac{t - 42.5}{35} \tag{7.21}$$

For the sake of clarity and simplification of the notation, we use the following designations: for the amount of sulphuric acid in the electrolyte $m(H_2SO_4) = a$; for the amount of aluminium in electrolyte $m(Al) = b$; for the electrolyte temperature $T = c$. Considering the Eq. (7.12*) and the transformation Eqs. (7.18)–(7.21), we can express the *objective function for minimizing the deposition time* for independent variables in nature unit in the following form:

$$t = f_0(a, b, c)_{th = th_p} \quad \rightarrow \quad \min \tag{7.22}$$

Using the adjustments, we get the following

$$\begin{aligned}\left(2.857 \cdot 10^{-3} t - 1.214\right) K &= 4.112 + 0.291\, b - 1.322 \cdot 10^{-2} a\, c - 0.220\, c \\ &+ 3.437 \cdot 10^{-7} a^3 - 5 \cdot 10^{-4} ab - 2.057 \cdot 10^{-2} a \\ &+ 4.925 \cdot 10^{-3} c^2 - 1.662 \cdot 10^{-3} a\, b \\ &+ 1.804 \cdot 10^{-4} a^2 + 7.554 \cdot 10^{-5} a + th_p\end{aligned} \tag{7.23}$$

where

$$K = -9.289 \cdot 10^{-5} a^2 + 3.251 \cdot 10^{-2} a + 1.205. \tag{7.24}$$

If in the relation (7.23) we denote the right-hand side PS, then for the *optimality criterion*, the deposition time t, we get a simplified form of the expression:

$$t = \left(\frac{PS}{K} + 1.214\right) \cdot \frac{10^3}{2.857}. \tag{7.25}$$

In the relation (7.25), K appears in fraction denominator, therefore we must consider the condition for K, i.e., $K \neq 0$. Taking this condition into account and using the Eq. (7.24), we come to the solution of the quadratic equation

$$-9.289 \cdot 10^{-5} a^2 + 3.251 \cdot 10^{-2} a + 1.205 = 0. \tag{7.26}$$

The Eq. (7.26) is satisfied by the two roots $a_1 = -33.807$, $a_2 = 383.807$. However, in engineering practice, we can only consider the positive root a_2, since the amount of sulphuric acid in the electrolyte cannot be a negative number. Using the above relations, we break down the Eq. (7.25).

The objective function $t = f_0(a, b, c)$ for *optimizing the deposition time of the AAO process* for independent variables in nature unit takes the following form after breaking it down:

$$t = f_0(a,b,c) = \frac{143.933 + 10.179\,b + 0.172\,c^2 - 1.209 \cdot 10^{-5}a^3}{-9.289 \cdot 10^{-5}a^2 + 3.251 \cdot 10^{-2}a + 1.205} + \frac{6.316 \cdot 10^{-3}\,a^2 - 7.706\,c - 0.462\,c\,b - 1.750 \cdot 10^{-2}\,c\,a}{-9.289 \cdot 10^{-5}a^2 + 3.251 \cdot 10^{-2}a + 1.205} + \frac{2.644.10^{-3}ab - 5.816 \cdot 10^{-2}ba - 0.72\,a + 35\,th_p}{-9.289 \cdot 10^{-5}a^2 + 3.251 \cdot 10^{-2}a + 1.205} + 42.5 \tag{7.27}$$

where

a is the amount of H_2SO_4 in electrolyte, in nature unit in $\left[g \cdot 1^{-1}\right]$.
b is the amount of Al in the electrolyte in nature unit in $\left[g \cdot 1^{-1}\right]$.
c electrolyte temperature, in nature unit in [°C].
t deposition time of the AAO layer in nature unit in [min].
th_p prescribed thickness of the formed AAO layer in [μm].

When defining the limiting conditions of AAO by the GS method, we must consider the proven intervals of technological constraints and at the same time consider the constraints that are given by the experiment, i.e., the data given in Tables 7.5, 7.9 and 7.10. Thus, we define technological constraints in the form of inequalities

$$165 \leq m(H_2SO_4) \leq 225 \tag{7.28}$$

$$1.5 \leq m(Al) \leq 9 \tag{7.29}$$

$$12 \leq T \leq 24 \tag{7.30}$$

To run the optimization procedure for solving the minimization problem with the objective function (7.27) and constraints (7.28)–(7.30), the appropriate script was created in MATLAB using nonlinear programming within the Optimization Toolbox. The goal of nonlinear optimization in MATLAB environment was to find the minimum (local extremum) of the objective function for the optimization problem written in a format suitable for the given software, i.e., in the following form:

$$\min_{x} f(x) \begin{cases} c(x) \leq 0 \\ ceq(x) = 0 \\ \mathbf{A} \cdot \mathbf{x} \leq \mathbf{b} \\ \mathbf{Aeq} \cdot \mathbf{x} = \mathbf{beq} \\ \mathbf{lb} \leq x \leq \mathbf{ub} \end{cases} \tag{7.6}$$

where **x**, **b**, **beq**, **lb** (lower boundary) and **ub** (upper boundary) are vectors, **A** and **Aeq** are matrices with constant coefficients, $c(x)$ and $ceq(x)$ are vector functions and $f(x)$ is a scalar function.

Functions $f(x)$, $c(x)$ and $ceq(x)$ re nonlinear [32]. The objective function (7.27) and the constraints (7.28)–(7.30) have been rewritten into a form suitable for optimization in MATLAB.

After implementation of the optimization procedure, we obtained the optimal combination of the values of the control variables and the optimal time for AAO using GS method. The results of the anodizing optimization for the individual required thicknesses of the AAO layer formed at the current density $J_A = 1\left[\mathrm{A \cdot dm^{-2}}\right]$ are given in Table 7.11. From Table 7.11 it follows that resulting deposition time is mainly influenced by the amount of sulphuric acid in the electrolyte. Thus, the optimal (minimum) value of the deposition time is achieved by increasing the amount of sulphuric acid in the electrolyte while maintaining constant values of the amount of aluminium in the electrolyte at the level $m(Al) = 9\,[\mathrm{g \cdot l^{-1}}]$ and the temperature of the electrolyte at the level $T = 24\,[\,^{\circ}\mathrm{C}]$. These results are consistent with the results of our previous experiments, as confirmed by Fig. 7.8. It is also necessary to consider the value of the critical temperature when the formed layer dissolves and layers of unsatisfactory quality are formed. The critical temperature value varies for optimization conditions at the level $T = 31\,[\,^{\circ}\mathrm{C}]$.

The graphical output of the anodic oxidation optimization for the desired layer thickness $th_p = 10\ [\mu\mathrm{m}]$ is presented in Fig. 7.9.

The graphical output of the optimization procedure for AAO with the required layer thickness $th_p = 30\ [\mu\mathrm{m}]$ is presented in Fig. 7.10.

Table 7.11 AAO process optimization results for at $J_A = 1[\mathrm{A \cdot dm^{-2}}]$

Required layer thickness [μm]	m (H_2SO_4) [$\mathrm{g \cdot l^{-1}}$]	m (Al) [$\mathrm{g \cdot l^{-1}}$]	T [°C]	$t_{optimum}$ [min]
10	165.806	9	24	48.0820
12	167.309	9	24	49.8132
14	168.435	9	24	51.5436
15	168.894	9	24	52.4086
20	170.515	9	24	56.7323
25	171.479	9	24	61.0548
30	172.109	9	24	65.3768

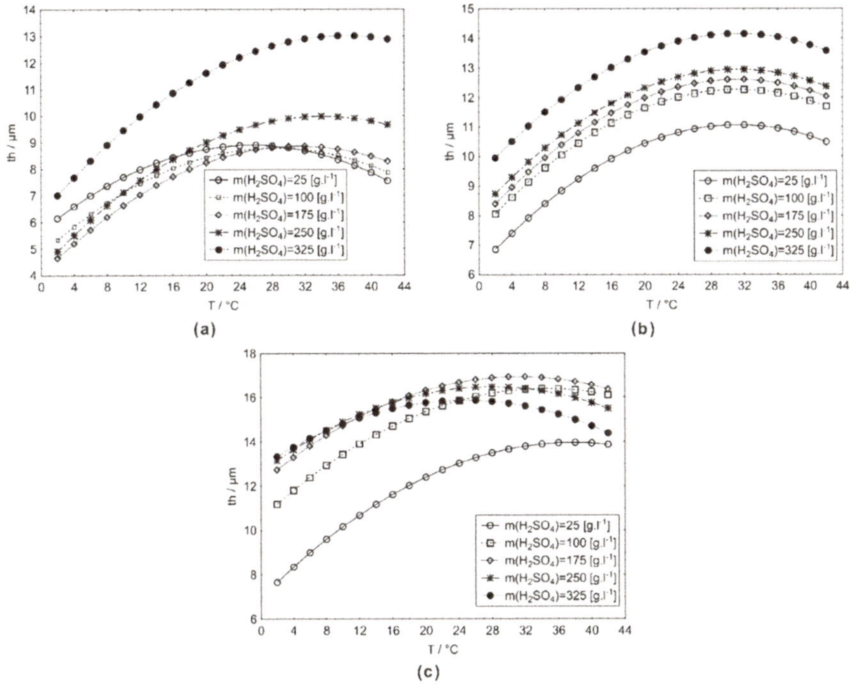

Fig. 7.8 Influence of electrolyte temperature on the thickness of the formed layer while changing the amount of sulphuric acid for $J_A = 1$ [A.dm^{-2}] **a** t = 10 [min], **b** t = 42.5 [min], **c** t = 80[min]

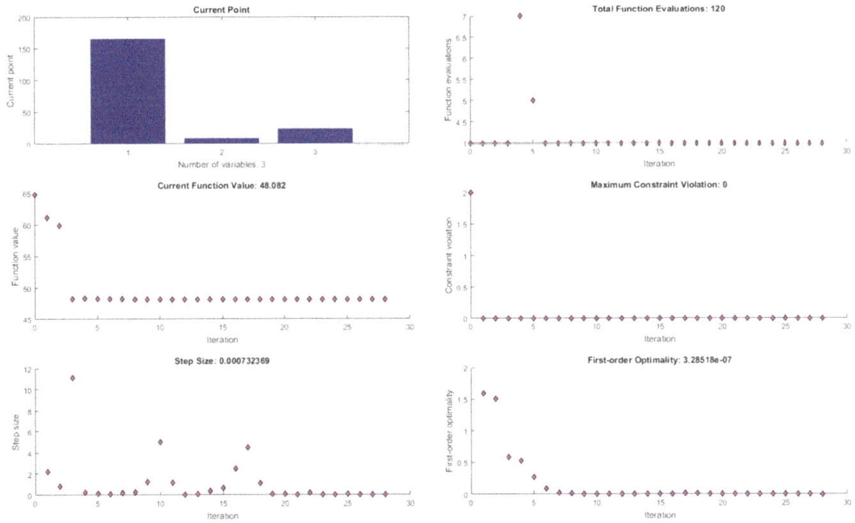

Fig. 7.9 Graphical presentation of outputs of the optimization process for the layer thickness 10 [µm]

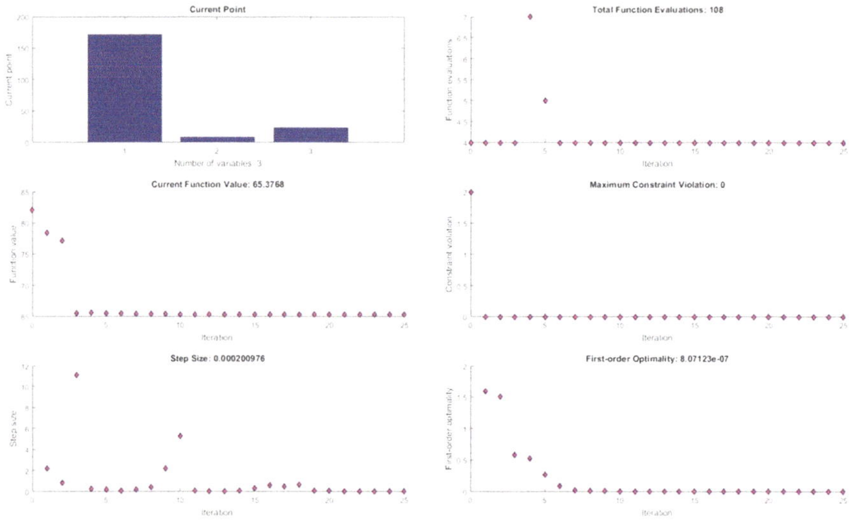

Fig. 7.10 Graphical presentation of outputs of the optimization process for the layer thickness 30 [μm]

Let us proceed in a similar way when optimizing AAO for the anode current density $J_A = 2\left[\mathrm{A}\cdot\mathrm{dm}^{-2}\right]$. We will follow from the Eq. (7.14) in which we simply denote h the left side, thus:

$$\begin{aligned} h = {} & 13.56 + 2.91\,x_4 + 4.94\,x_1 + 1.91\,x_3 - 1.32\,x_1x_3 \\ & - 2.19\,x_3^2 - 3.57\,x_4^2x_1 - 1.51\,x_1\,x_3\,x_4 \end{aligned} \tag{7.14*}$$

By adjusting deposition time x_4 (in coded unit) we can get the equation

$$\begin{aligned} 3.57\,x_4^2x_1 - (2.91 - 1.51\,x_1\,x_3)x_4 = {} & 13.56 + 4.94\,x_1 + 1.91\,x_3 \\ & - 1.32\,x_1x_3 - 2.19\,x_3^2 - h \end{aligned} \tag{7.31}$$

Having solved the Eq. (7.31), we get two solutions under the condition $x_1 \neq 0$ for deposition time x_4. Fulfilment of this condition shows that the amount of sulphuric acid in the electrolyte is different from zero and due to the real conditions of practice we are not interested in the solution for $x_1 = 0$, when optimizing the AAO process we are only interested in these solutions for $x_1 \neq 0$

$$\left\{\frac{0.0014(\sigma_1 - 151\,x_1x_3 + 291)}{x_1};\ -\frac{0.0014(151\,x_1x_3 + \sigma_1 - 291)}{x_1}\right\} \tag{7.32}$$

where

$$\sigma_1 = \sqrt{1.93 \cdot 10^7 x_1 + 2.28 \cdot 10^5 x_1^2 x_3^2 - 1.42 \cdot 10^6 h\, x_1 + 1.84 \cdot 10^6 x_1 x_3}$$
$$+ \sqrt{7.04 \cdot 10^6 x_1^2 - 1.87 \cdot 10^6 x_1^2 x_3 + 8.46 \cdot 10^5 - 3.12 \cdot 10^6 x_1 x_3^2} \quad (7.33)$$

Using the transformation relations (7.18)–(7.21) and relations (7.32) and (7.33) we can express the *objective function for minimizing the deposition time* for $J_A = 2\left[\mathrm{A \cdot dm^{-2}}\right]$ and independent variables in nature unit in the already known form (7.22):

$$t = f_0(a, b, c)_{th = th_p} \quad \rightarrow \quad \min \quad (7.22)$$

where the amount of sulphuric acid in the electrolyte: $m(H_2SO_4) = a$, the amount of aluminium in the electrolyte: $m(Al) = b$ and the temperature of the electrolyte: $T = c$. After the implementation of the optimization procedure, we obtained the optimal combination of the values of the control variables and the optimal time for AAO using the GS method and for the current density $J_A = 2\left[\mathrm{A \cdot dm^{-2}}\right]$. The results of anodizing optimization for the individual required thicknesses of the AAO layer formed at this current density are given in Table 7.12.

To run the optimization procedure in MATLAB, the m-file, which uses nonlinear programming methods in the Optimization Toolbox and considers the technological constraints (7.28)–(7.30) defined for AAO with the defined local current density, was created.

From the Table 7.12 it follows that by increasing the current density to $J_A = 2\left[\mathrm{A \cdot dm^{-2}}\right]$ we get an average time saving with an optimal setting of the operating factors by 35.39% as compared to the current density $J_A = 1\left[\mathrm{A \cdot dm^{-2}}\right]$. Even in this case, the most important criterion is the amount of sulphuric acid in the electrolyte. Thus, the optimal (minimum) value of the deposition time is again achieved by changing the amount of sulphuric acid in the electrolyte while maintaining constant values of the amount of aluminium in the electrolyte. Because the amount of aluminium in the electrolyte is not a significant factor at this current density, we consider its value at the lower limit: $m(Al) = 1,5\,[\mathrm{g \cdot l^{-1}}]$. We choose the temperature at the upper limit of the selected interval, as presented in Table 7.12.

Table 7.12 AAO process optimization results for $J_A = 2\left[\mathrm{A \cdot dm^{-2}}\right]$

Required layer thickness [μm]	m (H_2SO_4) [g·l^{-1}]	m (Al) [g·l^{-1}]	T [°C]	$t_{optimal}$ [min]
10	175.000	1.5	17.566	24.2561
12	176.235	1.5	17.116	29.2809
14	178.368	1.5	17.126	34.3058
15	179.544	1.5	17.131	36.8183
20	180.678	1.5	18.011	39.3805
25	182.112	1.5	17.190	41.9427
30	184.523	1.5	17.219	44.5049

7.2.4 Discussion of the Results of Mathematical Modelling and Optimization of the Anodic Aluminium Oxidation Process

In practice, it is necessary to find more comprehensive approach to the issue of optimization of surface treatment processes aa well as to consider critical values of some factors [43, 48]. In the AAO process we must also consider the value of critical temperature. When it is exceeded, the layer that is being formed dissolves or layers of unsatisfactory quality are formed. Therefore, to increase the efficiency of the AAO operation in practice, our research and experiments were focused on the analysis of the time of AAO layer formation and determination of critical temperature.

Anodic oxidation time is a significant factor influencing the thickness of the formed anodic layer at the considered current densities $J_A = 1\left[\mathrm{A}\cdot\mathrm{dm}^{-2}\right]$ and $J_A = 2\left[\mathrm{A}\cdot\mathrm{dm}^{-2}\right]$ used in practice. This input factor is manifested by its influence both independently and in interactions with other acting factors depending on the current density. Graphical representations of the dependence of the thickness of the formed anodic layer on anodic oxidation time for different settings of the electrolyte temperature and the amount of sulphuric acid in the electrolyte (due to the existence of statistically significant interactions with these factors) are shown in Figs. 7.11 and 7.12 for local anode current density $J_A = 1\left[\mathrm{A}\cdot\mathrm{dm}^{-2}\right]$ and $J_A = 2\left[\mathrm{A}\cdot\mathrm{dm}^{-2}\right]$.

At current density $J_A = 1\left[\mathrm{A}\cdot\mathrm{dm}^{-2}\right]$, anodic oxidation time is a significant factor influencing the thickness of the formed layer at the level of significance $\alpha = 5\,\%$. Its effect on the variability of the layer thickness is 23% and it is the most significant factor. Anodic oxidation time further occurs in statistically significant interactions with the amount of sulphuric acid in the electrolyte and its temperature. We attribute a 10% effect to this interaction on the change in layer thickness. Anodic oxidation time in a predictive mathematical and statistical calculation model is significant with the square of the amount of sulphuric acid while the effect of this interaction is 10%.

As we can see in Fig. 7.11, with minimal amounts of sulphuric acid in the electrolyte, i.e., $m(H_2SO_4) = 25\,[\mathrm{g}\cdot\mathrm{l}^{-1}]$, we observe an increase in the thickness of the coated layer under the influence of increasing anodic oxidation time over the entire range of electrolyte temperatures from $T = 2\,[^\circ\mathrm{C}]$ to $T = 42\,[^\circ\mathrm{C}]$. This increase in the thickness of the layer formed raises linearly with increasing temperature. At lower electrolyte temperatures, increase in the thickness of the layer due to increasing anodic oxidation time is about 19%. By increasing the electrolyte temperature to $T = 12\,[^\circ\mathrm{C}]$, increase in the thickness of the anodic layer is about 25%. As the temperature increases, the increase in layer thickness raises up to 45.5% at the electrolyte temperature $T = 42\,[^\circ\mathrm{C}]$. In the area of low electrolyte temperatures, the layer thickness reaches its maximum value at $T = 22\,[^\circ\mathrm{C}]$.

After substituting the experimental amount of sulphuric acid in the electrolyte, specifically $m(H_2SO_4) = 25\,[\mathrm{g}\cdot\mathrm{l}^{-1}]$, we can write the dependence of the thickness of the formed anodic layer for $J_A = 1\left[\mathrm{A}\cdot\mathrm{dm}^{-2}\right]$ into the relation (7.13) in the following form:

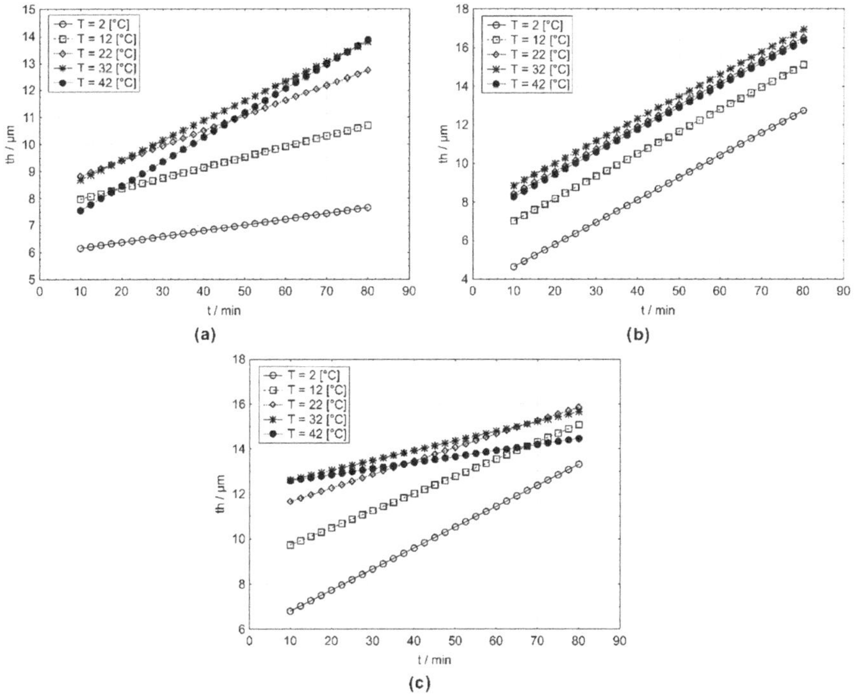

Fig. 7.11 Dependence of the thickness of the formed layer on anodic oxidation time at the current density $J_A = 1\ \left[\mathrm{A\cdot dm^{-2}}\right]$. Dependence of the thickness of the formed layer on anodic oxidation time at the current density with respect to the electrolyte temperature and the amount of sulphuric acid in the electrolyte: **a** $m(H_2SO_4) = 25\ \left[\mathrm{g\cdot l^{-1}}\right]$; **b** $m(H_2SO_4) = 175\ \left[\mathrm{g\cdot l^{-1}}\right]$; **c** $m(H_2SO_4) = 325\ \left[\mathrm{g\cdot l^{-1}}\right]$

$$th_{J_A=1\ [\mathrm{A\cdot dm^{-2}}];\ m(H_2SO_4)=25\ [\mathrm{g\cdot l^{-1}}]} = f(t,T) = 0.235\,T + 0.018\,t + 17.17\cdot 10^{-4}\,T\,t - 4.929\cdot 10^{-3}\,T^2 + 5.489 \tag{7.34}$$

The formula (7.34) is an experimentally obtained functional dependence expressing the thickness change of the formed layer depending on anodic oxidation time at a given electrolyte temperature. By the partial derivation (7.34) according to time, we get the theoretical speed of anodic layer formation at the amount $m(H_2SO_4) = 25\,[\mathrm{g\cdot l^{-1}}]$ in the electrolyte in the following form:

$$v_{th_{J_A=1\ [\mathrm{A\cdot dm^{-2}}];\ m(H_2SO_4)=25\ [\mathrm{g\cdot l^{-1}}]}} = \frac{\partial\, th_{J_A=1\ [\mathrm{A\cdot dm^{-2}}];\ m(H_2SO_4)=25\ [\mathrm{g\cdot l^{-1}}]}}{\partial\, t} = \frac{\partial f(t,T)}{\partial\, t} = 0.01815 + 0.001718\,T \tag{7.35}$$

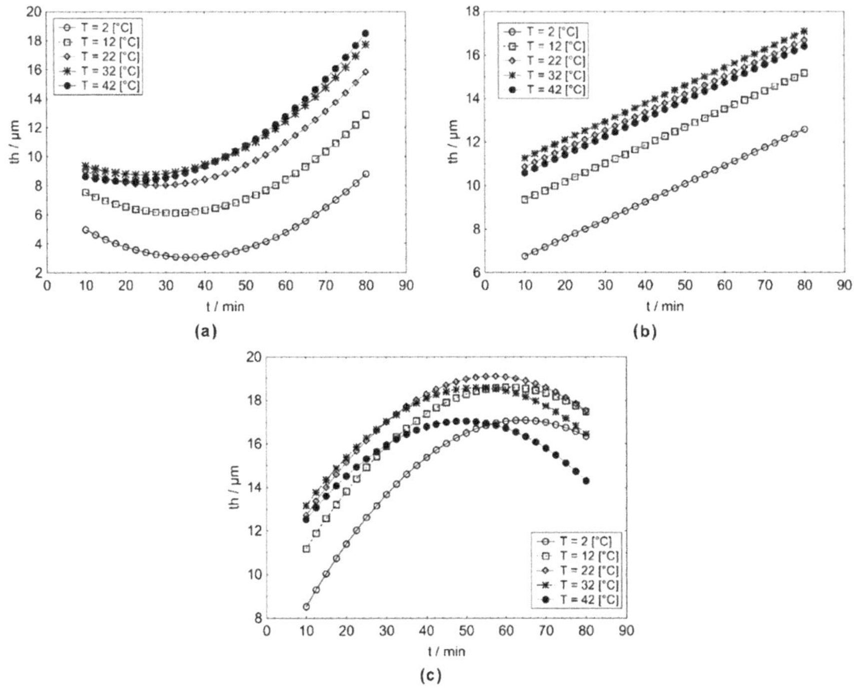

Fig. 7.12 Dependence of the thickness of the formed layer on anodic oxidation time at current density $J_A = 2\left[\mathrm{A}\cdot\mathrm{dm}^{-2}\right]$ with respect to the electrolyte temperature and the amount of sulphuric acid in the electrolyte: **a** $m(H_2SO_4) = 25\left[\mathrm{g}\cdot\mathrm{l}^{-1}\right]$; **b** $m(H_2SO_4) = 175\left[\mathrm{g}\cdot\mathrm{l}^{-1}\right]$; **c** $m(H_2SO_4) = 325\left[\mathrm{g}\cdot\mathrm{l}^{-1}\right]$

For a given amount of sulphuric acid in the electrolyte, the speed of layer formation is a function of the temperature of the electrolyte. Both from the graphical interpretation and the Eq. (7.35) it is evident that the speed of layer formation raises linearly with increasing of the electrolyte temperature. If we substitute analogously other values of the amount of sulphuric acid, i.e., $m(H_2SO_4) = 175\,[\mathrm{g}\cdot\mathrm{l}^{-1}]$ and $m(H_2SO_4) = 325\,[\mathrm{g}\cdot\mathrm{l}^{-1}]$ into the calculation Eq. (7.13) and again express the partial derivatives of the acquired functional dependences. We obtain the relations for the theoretical speed of layer formation for a given chemical composition of the electrolyte

$$v_{th_{J_A=1\,[\mathrm{A}\cdot\mathrm{dm}^{-2}];\,m(H_2SO_4)=25\,[\mathrm{g}\cdot\mathrm{l}^{-1}]}} = \frac{\partial\, th_{J_A=1\,[\mathrm{A}\cdot\mathrm{dm}^{-2}];\,m(H_2SO_4)=175\,[\mathrm{g}\cdot\mathrm{l}^{-1}]}}{\partial t} = \frac{\partial f(t,T)}{\partial t} = 0.115635 + 7.5\cdot 10^{-13}\,T \tag{7.36}$$

$$v_{th_{J_A=1\,[\mathrm{A}\cdot\mathrm{dm}^{-2}];\,m(H_2SO_4)=325\,[\mathrm{g}\cdot\mathrm{l}^{-1}]}} = \frac{\partial\, th_{J_A=1\,[\mathrm{A}\cdot\mathrm{dm}^{-2}];\,m(H_2SO_4)=325\,[\mathrm{g}\cdot\mathrm{l}^{-1}]}}{\partial t}$$

$$= \frac{\partial f(t,T)}{\partial t} = 0.09374 + 0.001718\ T \qquad (7.37)$$

Graphical representation of the dependence of the theoretical layer formation speed on the electrolyte temperature for a given amount of sulphuric acid (factor x_1) in the electrolyte is shown in Fig. 7.13. From the course of the theoretical speed of layer formation it follows that in addition to the electrolyte temperature, the speed is also affected by the amount of sulphuric acid in the electrolyte. With $m(H_2SO_4) = 25\,[\mathrm{g} \cdot \mathrm{l}^{-1}]$ the speed of layer formation increases with increasing electrolyte temperature. Conversely, at high concentrations of sulphuric acid, this speed decreases due to temperature, and approximately in the middle of the range of sulphuric acid used, the temperature of the electrolyte will not affect the speed of layer formation. For a comprehensive picture of the theoretical speed of layer formation we must logically consider the effect of the amount of sulphuric acid in the electrolyte.

By expressing the partial derivative of the function (7.13) with respect to time, we can obtain a more complex form for the dependence of the layer formation speed on the electrolyte temperature and the amount of sulphuric acid in the electrolyte for current density $J_A = 1\left[\mathrm{A} \cdot \mathrm{dm}^{-2}\right]$ in nature unit in the form (7.38) as it is shown in Fig. 7.14.

$$v_{th_{J_A=1\,[\mathrm{A}\cdot\mathrm{dm}^{-2}]}} = -0.009696 - 2.65273 \cdot 10^{-6}[m(H_2SO_4)]^2 + m(H_2SO_4)\left[11.8041 \cdot 10^{-4} - 11.4524 \cdot 10^{-6}\,T\right] + 2.00417 \cdot 10^{-3}\,T \quad (7.38)$$

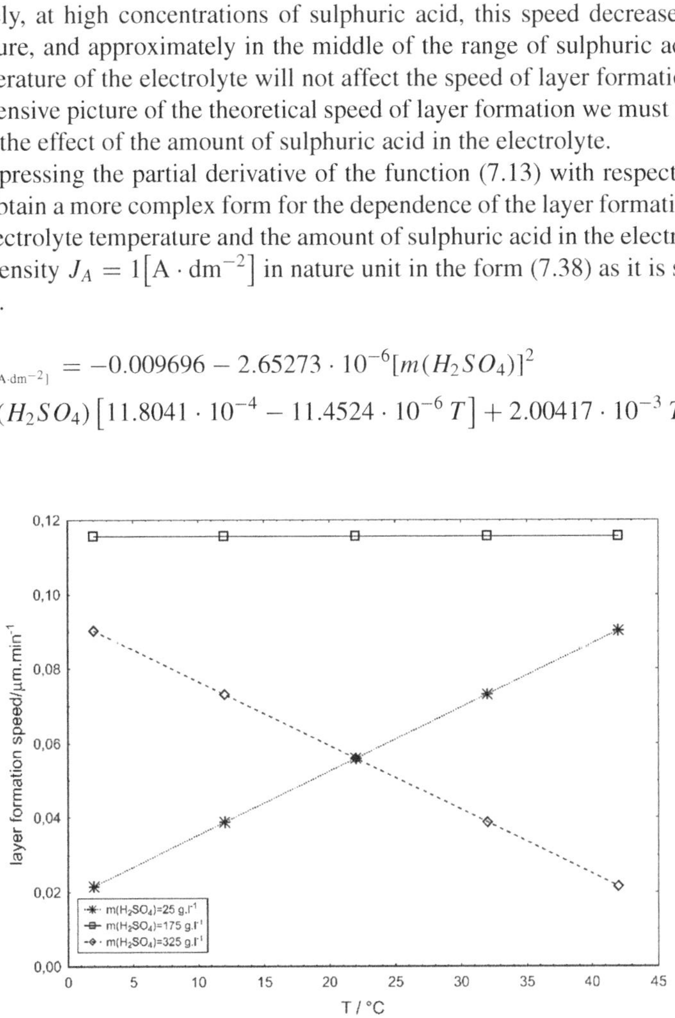

Fig. 7.13 The course of the theoretical speed of layer formation depending on the considered change in the temperature of the electrolyte at a given amount of sulphuric acid

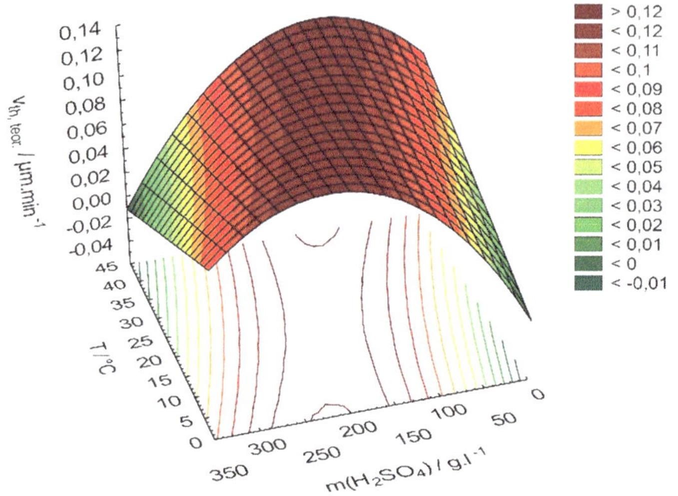

Fig. 7.14 Dependence of the theoretical speed of layer formation on the electrolyte temperature and amount of $m(H_2SO_4)$ in electrolyte at current density of $J_A = 1\left[\text{A} \cdot \text{dm}^{-2}\right]$

The graph shows that theoretical speed of anodic layer formation decreases with the increasing electrolyte temperature and increasing amount of sulphuric acid in the electrolyte. The speed of layer formation also decreases around low electrolyte temperatures and low concentrations of sulphuric acid in the electrolyte.

When analysing the effect of the electrolyte temperature on the formation of the coated layer in the anodic aluminium oxidation, the factor x_3 appears to be significant at the level of significance $\alpha = 5\,\%$ both as a major effect and in the form of interactions, especially with the amount of sulphuric acid in the electrolyte and in combination with anodic oxidation time. The importance of this factor lies primarily in influencing the conductivity of the electrolyte and thus accelerating electrochemical reactions. The influence of the electrolyte temperature on the thickness of the formed AAO layer at the current density $J_A = 1\left[\text{A} \cdot \text{dm}^{-2}\right]$ is presented in Fig. 7.15.

According to Fig. 7.15 it is evident that for each deposition time setting, i.e., $t = 10$[min], $t = 42.5$[min], $t = 80$[min] and each adjustment of the amount of sulphuric acid in the electrolyte, the effect of the electrolyte temperature on the thickness of just formed layer can be divided into two areas, namely into two intervals for the temperature T of the electrolyte.

The first area is characterized by the fact that with increasing temperature, the thickness th of the just formed layer increases while the for the temperature T from the second interval this dependence no longer applies. On the contrary, with further temperature increase, the thickness of the formed layer decreases due to its dissolution. By determining the local extremum for temperature T for the obtained functional dependences (graphically presented in Fig. 7.15), the values of the so-called critical electrolyte temperature $T_{krit.}$ have been set. For example, during the

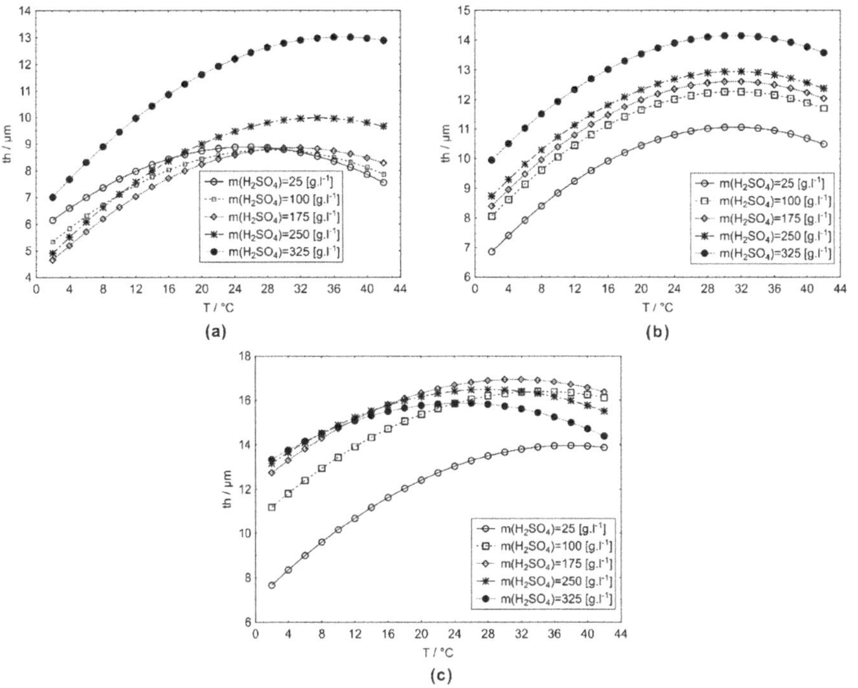

Fig. 7.15 Dependence of the thickness of the formed AAO layer on the electrolyte temperature at current density $J_A = 1\left[\mathrm{A}\cdot\mathrm{dm}^{-2}\right]$ regarding the amount of sulphuric acid in the electrolyte and anodizing time: **a** $t = 10$ [min]; **b** $t = 42.5$ [min]; **c** $t = 80$ [min]

AAO process $t = 10\,[\mathrm{min}]$ at current density $J_A = 1\left[\mathrm{A}\cdot\mathrm{dm}^{-2}\right]$ and setting the sulphuric acid concentration to $m(H_2SO_4) = 25\,[\mathrm{g}\cdot\mathrm{l}^{-1}]$ the critical temperature $T_{krit.} = 26\,[\,°\mathrm{C}]$ was set (see Fig. 7.15a). In such a case, the setting of the anodizing process conditions is the first area (the area of increasing thickness of the AAO layer) for $T \in \langle 2,\ 26)$, the second area (the area of decreasing the thickness and decreasing the quality of the AAO layer) is for $T > 26\,[\,°\mathrm{C}]$. The increase in thickness in the first area is about 30%. Increase in the temperature of the electrolyte above 26 [°C] causes dissolution of the formed layer, which is reflected in a decrease in the thickness and quality of the formed coating. The decrease in thickness in the second area represents 17.6%.

When setting the anodizing process conditions to the values $t = 42.5\,[\mathrm{min}]$, $J_A = 1\left[\mathrm{A}\cdot\mathrm{dm}^{-2}\right]$, $m(H_2SO_4) = 25\,[\mathrm{g}\cdot\mathrm{l}^{-1}]$, then the critical temperature of the value $T_{krit.} = 31\,[\,°\mathrm{C}]$ was determined. For $T \in \langle 2;\ 31)$, the increase in the thickness of the just formed layer will increase to about 38% with increasing temperature. At temperatures above $T_{krit.} = 31\,[\,°\mathrm{C}]$, the layer thickness decreases again, in this case only 5%. With this setting of the anodizing duration $t = 42.5\,[\mathrm{min}]$, we can see from Fig. 7.15b that further increase of the concentration of sulphuric acid in the electrolyte will not cause a shift of the maximum value of the thickness th [µm] of the formed layer depending on the temperature. Local extremum—maximum for the thickness of the formed AAO layer occurs at temperature $T_{krit.} = 31\,[\,°\mathrm{C}]$. The absolute value of the thickness of the formed layer varies and the change in layer thickness at this critical electrolyte temperature is 21% with an increase in the amount of sulphuric acid in the electrolyte from 25 $[\mathrm{g}\cdot\mathrm{l}^{-1}]$ to 325 $[\mathrm{g}\cdot\mathrm{l}^{-1}]$. During the AAO process $t = 80\,[\mathrm{min}]$, we obtained for individual concentrations $m(H_2SO_4)$ in the electrolyte critical temperature values ranging from $T_{krit.} = 37.74\,[\,°\mathrm{C}]$ for $m(H_2SO_4) = 25\,[\mathrm{g}\cdot\mathrm{l}^{-1}]$ to $T_{krit.} = 24.68\,[\,°\mathrm{C}]$ for the concentration of sulphuric acid $m(H_2SO_4) = 325\,[\mathrm{g}\cdot\mathrm{l}^{-1}]$ (see Fig. 7.15c). Analogously, graphical dependences were created for the AAO process realized at the current density $J_A = 2\left[\mathrm{A}\cdot\mathrm{dm}^{-2}\right]$ and the values $T_{krit.}$ of the critical electrolyte temperature were also determined.

Since in scientific and professional works of many authors we encounter the approach of evaluating the influence of only one selected factor (so-called COST approach) on a selected parameter of the layer formed during anodic aluminium oxidation and galvanizing, it was desirable to analyse these technological processes more comprehensively. Surface treatments as well as engineering and electrochemical processes represent complex, multifactor and often nonlinear systems, where the mutual interactions of the acting factors also play an important role. Therefore, an approach where we can simultaneously monitor the mutual influence of several acting factors is the closest to the real conditions of the processes. Such a real approximation contributes to a more accurate simulation of a given technological process and its deeper understanding and subsequent optimization. Application of mathematical-statistical methods in the evaluation of experimentally obtained data in the process of anodic aluminium oxidation allowed us in conjunction with application of the DOE methodology not only to identify significant factors and their interactions that affect the change of the monitored parameter, i.e., the thickness of the formed layer, but also to compile computational mathematical models predicting the thickness of the formed layer and to determine the value of the critical temperature. This has a significant benefit for practice because based on practical verification of the developed models in the conditions of real production, it has been shown that by applying these models it is possible to create a protective layer with the thickness required by customers.

References

1. Box, G. E. P., Hunter, J. S., & Hunter, W. G. (2005). *Statistics for experimenters* (2nd ed., 639p.). John Wiley & Sons, Inc. ISBN 978-0-471-71813-0.
2. Eriksson, L., Johansson, E., Kettaneh-Wold, N., Trygg, J., Wikström, C., & Wold, S. *Multi-and megavariate data analysis. Part I basic principles and applications* (425p.). Umetrics Academy. ISBN 978-91-973730-2-8.
3. Eriksson, L., Johansson, E., Kettaneh-Wold, N., Trygg, J., Wikström, C., & Wold, S. (2006). *Multi-and megavariate data analysis. Part II basic principles and applications* (307p.). Umetrics Academy. ISBN 978-91-973730-2-9.

4. Panda, A., Jurko, J., Džupon, M., & Pandová, I. (2011). Optimalization of heat treatment bearings rings with goal to eliminate deformation of material. In *Chemické listy, Material in Engineering Practice 2011*, (Vol. 105, pp. 459–461), Herľany, Special issue, Asociation of Czech chemical society Praha, HF TU Košice, ISSN 0009-2770.
5. Panda, A., Duplák, J., Jurko, J., & Behún, M. (2013, November 24–25). New experimental expression of durability dependence for ceramic cutting tool. In *Applied mechanics and materials, ICAMM 2012, international conference on applied mechanics and materials* (Vol. 275–277, pp. 2230–2236), Sanya, China. ISBN 978-303785591-1, ISSN 1660-9336.
6. Hendl, J. a kol. (2014). *Statistika v aplikacích* (456p.). Portál. ISBN 978-80-262-0700-9.
7. Meloun, M., & Militký, J. (2006). *Kompendium statistického zpracování dat* (982p.). 2. vyd. Academia. ISBN 80–200–1396–2.
8. Meloun, M., & Militký, J. (2004). *Statistická analýza experimentálnich dat* (953p.). Academia. ISBN 80-200-1254-0.
9. Meloun, M., Militký, J., & Hill, M. (2005). *Počítačová analýza vícerozměrných dat v příkladech* (449p.). Academia. ISBN 80-200-1335-0.
10. Michal, P., Piteľ, J., Vagaská, A., & Bukovský, I. (2014, 6–11 July). Application of neural networks to evaluate experimental data of galvanic zincing. In *Proceedings of the 2014 International Joint Conference on Neural Networks (IJCNN)* (pp. 2997–3001), Beijing, China. IEEE. https://doi.org/10.1109/IJCNN.2014.6889799.
11. Boshkov, N., Petrov, K., Kovacheva, D., Vitkova, S., & Nemska, S. (2005). Influence of the alloying component on the protective ability of some zinc galvanic coatings. *Electrochimica Acta, 51*(1), 77–84. https://doi.org/10.1016/j.electacta.2005.03.049
12. Dubent S., Mertens, M. L. A. D., & Saurat M. (2010). Electrodeposition, characterization and corrosion behaviour of tin–20 wt.% zinc coatings electroplated from a non-cyanide alkaline bath. *Materials Chemistry and Physics, 120*, 371–380. https://doi.org/10.1016/j.matchemphys.2009.11.017.
13. Lee, J. M. (2006). Numerical analysis of galvanic corrosion of Zn/Fe interface beneath a thin electrolyte. *Electrochimica Acta., 51*(16), 3256–3260. https://doi.org/10.1016/j.electacta.2005.09.026
14. Song, G. M., Sloof, W. G., Pei, Y. T., & De Hosson, J. Th. M. (2006). Interface fracture behavior of zinc coatings on steel. Experiments and finite element calculations. *Surface and Coatings Technology, 201*(7), 4311–4316.
15. Yadav, A. P., Katazama, H., Noda, K., Masuda, H., Nishikata, A., & Tsuru, T. (2007). Surface potential distribution over a zinc/steel galvanic couple corroding under thin layer of electrolyte. *Electrochimica Acta., 52*(9), 3121–3129.
16. Michal, P., Gombár, M., Vagaská, A., Piteľ, J., & Kmec, J. (2013). Experimental study and modeling of the zinc coating thickness. *Advanced Materials Research, 712–715*, 382–386. https://doi.org/10.4028/www.scientific.net/amr.712-715.382
17. Vagaská, A. (2023). Mathematical-statistical nonlinear model of zincing process and strategy for determining the optimal process conditions. *Mathematics, 11*, 771. https://doi.org/10.3390/math11030771
18. Vagaská, A., Gombár, M., & Kmec, J. (2012). Optimalizácia hrúbky zinkového povlaku vytvoreného galvanickým zinkovaním. *Strojárstvo extra*, č. *5*, s. 26/1–26/5. ISSN 1335-2938.
19. Vagaská, A., Gombár, M., Michal, P., Kmec, J., Badida, M., & Sobotová, L. (2013, June 25–27). Study of physical and chemical factors affecting the thickness of deposited zinc coating. In *ICET 2013: proceedings of international congress on engineering and technology* (pp. 116–121). Dubrovnik, Croatia. Rise Association. ISBN 978-80-87670-08-8.
20. Vagaská, A., Gombár, M., Kmec, J., & Michal, P. (2013). Statistical analysis of the factors effect on the zinc coating thickness. *Applied Mechanics and Materials, 378*, 184–189. https://doi.org/10.4028/www.scientific.net/amm.378.184
21. Alfantazi, A. M., & Dreisinger, D. (2001). The role of zinc and sulfuric acid concentrations on zinc electrowinning from industrial sulfate based electrolyte. *Journal of Applied Electrochemistry, 31*, 641–646. https://doi.org/10.1023/A:1017569400346

22. Aliofkhazraei, M., Alamdari, E. K., Zamanzade, M., Salasi, M., Behrouzghaemi, S., Heydari, J., Haghshenas, D. F., & Zolala, V. (2007). Empirical equations for electrical conductivity and density of Zn, Cd and Mn sulphate solutions in the range of electrowinning and electrorefining electrolytes. *Journal of Materials Science, 42*, 9622–9631. https://doi.org/10.1007/s10853-007-1971-5
23. Dhak, D., Mahon, M., Asselin, E., & Alfantazi, A. (2012). Characterizing industrially electrowon sticky zinc deposits. *Hydrometallurgy, 111*, 136–140. https://doi.org/10.1016/j.hydromet.2011.09.007
24. Xia, X., Zhitomirsky, I., & McDermid, J. R. (2009). Electrodeposition of zinc and composite zinc–yttria stabilized zirconia coatings. *Journal of Materials Processing Technology, 209*(2009), 2632–2640.
25. Yu, J., Wang, L., Su, L., Ai, X., & Yang, H. Temperature effects on the electrodeposition of zinc. *Journal of Electrochemical Society, 150*, C19–C23. https://doi.org/10.1149/1.1525269.
26. Morávka, J., Maroš, B., & Michalek, K. (2008). Vliv neortogonality plánu experimentu na statistickou korektnost modelu. In *Mezinárodní konference Technical Computing*, str. 73. http://dsp.vscht.cz/konference_matlab/MATLAB08.
27. Stewart, J., & Clegg, D. (2012) *Brief applied calculus* (491p.). BROOKS/COLE Cengage Learning. ISBN 978-1-111-57005-7.
28. Badida, M., Gombár, M., Mašlejová, A., Sobotová, L., Kmec, J., & Vagaská, A. (2015). Evaluation of zinc coating quality by statistical methods. *Przemysl Chemiczny, 94*(12), 2146–2149. https://doi.org/10.15199/62.2015.12.11.
29. Vagaská, A., Gombár, M., & Korauš, A. (2022). Mathematical modeling and nonlinear optimization in determining the minimum risk of legalization of income from criminal activities in the context of EU member countries. *Mathematics, 10*, 4681. https://doi.org/10.3390/math10244681
30. Vagaská, A., Gombár, M., & Straka, Ľ. (2022). Selected Mathematical optimization methods for solving problems of engineering practice. *Energies, 15*(6), 2205. https://doi.org/10.3390/en15062205
31. Hebák, P., & Hustopecký, J. (1987). *Vícerozměrné statistické metody s aplikacemi* (456p.). SNTL/ALFA. ISBN 80-01-01076-7.
32. Dybvig, P. H. (2012). Numerical methods for optimization. In *Fin500J mathematical foundations in finance* (25pp.).
33. Antoniou, A., & Lu, W. S. (2007). *Practical optimization. algorithms and engineering applications* (675p.). Springer Science & Business Media LCC. ISBN-13: 978-0-387-71106-5.
34. Djordievic, S. S. (2020). Some unconstrained optimization methods. In *Applied Mathematics*. Intechopen books. https://doi.org/10.5772/intechopen.83679.
35. Vasilko, K. (2009). *Teória a prax trieskového obrábania* (532p.). FVT TU v Košiciach, Prešov. ISBN 978-80-553-0152-5.
36. Bocchetta, P., Sunseri, C., Chiavarotti, G., & Di Quarto, F. (2003). Microporous alumina membranes electrochemically grown. *Electrochimica Acta, 48*(20–22), 3175–3183. https://doi.org/10.1016/S0013-4686(03)00348-7
37. Garcia-Vergara, S. J., Skeldon, P., Thompson, G. E., & Habazaki, H. (2006). A flow model of porous anodic film growth on aluminium. *Electrochimica Acta, 52*(2), 681–687. https://doi.org/10.1016/j.electacta.2006.05.054.
38. Le Coz, F., Arurault, L., & Datas, L. (2010). Chemical analysis of a single basic cell of porous anodic aluminium oxide templates. *Materials Characterization, 61*(3), 283–288. https://doi.org/10.1016/j.matchar.2009.12.008, ISSN 1044–5803.
39. Panda, A., Dupľák, J., Jurko, J., & Behún, M. (2011). Comprehensive identification of sintered carbide durability in machining process of bearings steel 100CrMn6. In *Advanced Materials Research*, ICMPTMIT 2011 (Vol. 340, pp. 30–33). ISBN 978-3-03785-253-8, ISSN 1022–6680.
40. Sheasby, P. G., & Pinner, R. (2001). *The surface treatment and finishing of aluminium and its alloys* (6th ed., 743p.). ASM International, USA/Finishing Publications Ltd. ISBN 0-904477-21-5.

41. Thompson, G. E. (1997). Porous anodic alumina: Fabrication, characterization and applications. *Thin Solid Films, 297*(1–2), 192–201.
42. Gombár, M., Kmec, J., Badida, M., Sobotová, L., Vagaská, A., & Badidová, A. (2016). Analysis of physical factors on chosen properties of anodic alumina oxide (aao) layers and environment. *Metalurgija, 55*(4), 799–802.
43. Gombár, M., Vagaská, A., Harničárová, M., Valíček, J., Kušnerová, M., Czán, A., & Kmec, J. (2019). Experimental analysis of the influence of factors acting on the layer thickness formed by anodic oxidation of aluminium. *Coatings, 9*(1), 57. https://doi.org/10.3390/coatings9010057
44. Michal, P., Vagaská, A., Fechová, E., Gombár, M., & Kozak, D. (2016). Effect of electrolyte temperature on the thickness of anodic aluminium oxide (AAO) layer. *Metalurgija, 55*(3), 403–406. https://hrcak.srce.hr/153631.
45. Michal, P., Vagaská, A., Gombár, M., Kmec, J., Spišák, E., & Badida, M. (2014). Prediction of the effect of chemical composition of electrolyte on the thickness of anodic aluminium oxide layer. *International Journal of Mathematical Models and Methods in Applied Sciences, 8*(1), 152–155. ISSN: 1998-0140.
46. Spišák, E., Gombár, M., Kmec, J., Vagaská, A., Fechová, E., Michal, P., Piteľ, J., & Kučerka, D. (2015). Effect of the electrolyte temperature and the current density on a layer microhardness generated by the anodic aluminium oxidation. *Advances in Materials Science and Engineering, 2015*, Article ID 659846, 9p. https://doi.org/10.1155/2015/659846.
47. Vagaská, A., & Gombár, M. (2021). Mathematical optimization and application of nonlinear programming. In Š. Hošková-Mayerová, C. Flaut, & F. Maturo (Eds.), *Algorithms as a basis of modern applied mathematics*. Studies in fuzziness and soft computing (Vol. 404). Springer. https://doi.org/10.1007/978-3-030-61334-1_24.
48. Vagaská, A., & Gombár, M. (2019). Contribution to the application of neural networks to control the process of anodic oxidation of aluminium. In *18th Conference on Applied Mathematics, APLIMAT 2019* (Vol. 2, pp. 1227–1236). ISBN 978-151088214-0.
49. Vagaská, A., Michal, P., Bukovský, I., Gombár, M., & Kmec, J. (2014). Mathematical modelling and description of the technological process of aluminum anodic oxidation by using the neural networks. *International Journal of Materials, 1*, 9–16. https://www.naun.org/main/NAUN/materials/2014/a042018-099.pdf, ISSN 2313-0555.
50. Aerts, T., Jorcin, J. B., De Graeve, I., & Terryn, H. (2010). Comparison between the influence of applied electrode and electrolyte temperatures on porous anodizing of aluminium. *Electrochimica Acta, 55*(12), 3957–3965. https://doi.org/10.1016/j.electacta.2010.02.044
51. Aerts, T., Dimogerontakis, T., De Graeve, I., Fransaer, J., & Terryn, H. (2007). Influence of the anodizing temperature on the porosity and the mechanical properties of the porous anodic oxide film. *Surface and Coatings Technology, 201*(16–17), 7310–7317. https://doi.org/10.1016/j.surfcoat.2007.01.044
52. Djozan, D. J., & Amir-Zehni, M. (2003). Anodizing of inner surface of long and small-bore aluminum tube. *Surface & Coatings Technology, 173*(2–3), 85–191. https://doi.org/10.1016/S0257-8972(03)00510-3
53. Fratila-Apachitei, L. E., Duszczyk, J., & Katgerman, L. (2003). Vickers microhardness of AlSi(Cu) anodic oxide layers formed in H_2SO_4 at low temperature. *Surface & Coatings Technology, 165*(3), 309–315. https://doi.org/10.1016/S0257-8972(02)00750-8
54. Rahimi, M. H., Saramad, S., Tabaian, S. H., Marashi, S. P., Zolfaghari, A., & Mohammadalinezhad, M. (2009). (2009) Study the effect of striping in two-step anodizing process on pore arrangement of nano-porous alumina. *Applied Surface Science, 256*(1), 12–16.
55. Vojkuvka, L., Santos, A., Pallarès, J., Ferré-Borrull, J., Marsal, L. F., Celis, J. P. (2012) On the mechanical properties of nanoporous anodized alumina by nanoindentation and sliding tests. *Surface & Coatings Technology*. 206 (8–9), 2012, pp. 2115–2124.
56. Vagaská, A., Fechová, E., Michal, P., & Gombár, M. (2016). The influence of input factors of aluminium anodizing process on resulting thickness and quality of aluminium oxide layer, *Procedia Engineering, 149*, 512–519. https://doi.org/10.1016/j.proeng.2016.06.699, ISSN 1877-7058.

57. Vagaská, A., & Gombár, M. (2017). Comparison of usage of different neural structures to predict AAO layer thickness|Usporedba primjene raznih neuralnih struktura u predviđanju debljine sloja anodnog aluminij oksida. *Tehnicki Vjesnik, 24*(2), 333–339. https://doi.org/10.17559/TV-20140423164817.
58. Akolkar, R., Landau, U., Kuo, H., & Wang, Y. M. (2004). Modeling of the current distribution in aluminum anodization. *Journal of Applied Electrochemistry, 34*(2004), 807–813. https://doi.org/10.1023/B:JACH.0000035611.87036.36
59. Badida, M., Gombár, M., Sobotová, L., Kmec, J., Vagaská, A., & Michal, P. (2013, August 3–4). The influence of sodium chloride on the resulting AAO film thickness. *Advanced Materials Research, 816–817*, 18–22. *ICMST 2013, 4nd international conference on materials science and technology, Dubai, United Arab Emirates*. http://www.scientific.net/AMR, ISBN 978-303785867-7, ISSN 1662–8985.
60. Gombár, M., Kmec, J., Badida, M., Sobotová, L., Vagaská, A., & Michal, P. (2014). The simulation of the temperature effects on the microhardness of anodic alumina oxide layers. *Metallurgy, 53*(2014), 59–61.
61. Gombár, M., Vagaská, A., Kmec, J., & Michal, P. (2013, February). Microhardness of the coatings created by anodic oxidation of aluminium. *Applied Mechanics and Materials, 308*, 95–100. *Operation and diagnostics of machines and production systems operational states*. Trans Tech Publications, Ltd. https://doi.org/10.4028/www.scientific.net/amm.308.95. ISSN 1660-9336I, ISBN 978-3-03785-656-7.
62. Michal, P., Vagaská, A., Gombár, M., & Kmec, J. (2014). Mathematical modelling and optimization of technological process using design of experiments methodology. *Applied Mechanics and Materials, 616*(2014), 61–68. https://doi.org/10.4028/www.scientific.net/AMM.616.61
63. Michal, P., Vagaská, A., Gombár, M., Kmec, J., Spišák, E., & Kučerka, D. (2015). Usage of neural network to predict aluminium oxide layer thickness. *The Scientific World Journal*, Article number 253568, 10p. https://doi.org/10.1155/2015/253568.

Chapter 8
Conclusion

We have defined the space for the selected methods of mathematical programming and optimization. Our attention was focused on optimization and mathematical modelling related with the problem of choosing the optimal input conditions for technological processes. Specifically, we were interested in optimization of cutting conditions in turning and optimization of input variables during the galvanizing process and the process of anodic aluminium oxidation. We have presented application of linear, convex, and generally nonlinear methods of programming using the selected examples and issues arising in engineering activities. At the same time, we numerically and graphically illustrate the possibilities of applying software for optimization in MATLAB environment. As an optimality criterion, we have set the minimization of input costs and maximization of economic returns and profits.

A characteristic feature of the production processes of this time is the effort to produce at low cost but high quality in a short time. Because the required effort can be expressed as a function of certain decision-maker variables, the correct selection of input parameters of each process, also including technological processes, has a positive effect on output within the meaning of the above [1]. To find the right combination of values for the adjustable input variables of a given process has been the effort of experts for many years [2], [3]. It has been proved by research conducted for many years of experimenting. However, experimenting is often expensive and time consuming, as it is presented in the case with experiments of technological processes improvement. Especially, if the so-called COST approach is used when considering the influence of one selected factor on monitored response. That is when looking for the right combination for setting the input variables of a certain production, technological or any other process, it is important to take advantage of the possibilities offered by modern times: methods of modern applied mathematics, the DOE methods, statistical methods and in particular optimization methods in connection with progress in the field of computer technology and in development of special software [4], [5].

A. Vagaská et al., *Optimization Methods in Mathematical Modeling of Technological Processes*, Mathematical Engineering, https://doi.org/10.1007/978-3-031-35339-0_8

It is important to combine experimentation with optimization. It must go hand in hand. We cannot optimize without experimenting (we would not have mathematical model) and we cannot experiment without trying to optimize.

In the theoretical part of the monograph (Chaps. 1–4) we have acquainted with the theoretical mathematical foundations of optimization. Chapter 5 focused on optimal decision-making regarding control and optimize technological processes. Chapters 6–7 are of application nature. Based on the issues arising from real conditions of practice, we clarify the usefulness of the selected methods of applied mathematics, especially the usefulness of mathematical applications, statistical and optimization methods in experimental research and in the analysis of experimentally obtained data to "improve behaviour of the system", i.e., to optimize the operation of technological processes in engineering activities. The stated goal to "connect mathematical optimization with computer software for its application in selected technological processes optimization" has been fulfilled.

Machining processes and surface treatment processes were in the centre of our interest. We were interested in feed turning where we focused on optimizing its cutting conditions regarding the established criterion of optimality taking into consideration certain technological constraints. Because each combination of cutting parameters results in different roughness of the machined surface as well as tool lifespan [2], it is not easy to determine the best combination of cutting values parameters which provide a low roughness of the machined surface and the highest tool lifespan while ensuring minimum production costs. Several facts must be considered in machining processes. We should take into consideration apparent conflicts: if we increase the cutting speed, tool life will be reduced. When determining various factors for optimal cutting conditions, it is advantageous to use linear and nonlinear programming methods in MATLAB environment and implement them on turning process to minimize production costs as it was pointed out in Chap. 6.

In case of turning, we have considered the most common requirements of manufacturers and thus minimization of production costs per unit of production was chosen as the criterion of optimality. Additional requirements were focused on minimizing the power consumption of a cutting process or maximizing productivity while respecting certain constraints. Mathematical model (input for the optimization procedure) was based on already known empirical relations derived for the turning process. We have selected the shape suitable for application of a mathematical apparatus for linearization which made it possible to effectively solve the formulated optimization problem using techniques for linear programming (graphically and numerically in MATLAB).

Current trends in machine parts production are shifting to the area of maximum material savings with a constant increase of the accuracy of final dimensions of the components. The field of materials surface treatment must also adapt this effort. Requirements for mechanical and decorative properties of the created layers and their ability to protect surface-treated elements against corrosion while ensuring required thickness and tolerance of functional dimensions force us to look at the layers from both scientific and technological point of view to ensure all the needs and requirements of end customers.

In Chap. 7 we present the possibilities of applying the design of experiments (DOE) methodology as a scientifically justifiable methodology of cost and time-saving experimentation [1]. Then we present the possibilities of optimizing surface processes modifications applying nonlinear programming using Optimization Toolbox in MATLAB. We focused on optimizing the galvanizing and the anodic aluminium oxidation processes respecting the chosen optimality criterion. When analysing technological process of galvanizing and anodic aluminium oxidation, it is important to monitor the interaction of several factors at the same time. It is more appropriate to experiment with the DOE approach, which is illustrated when solving specific practical issues.

In case of surface treatment processes (galvanizing, anodic aluminium oxidation), we have developed mathematical models based on our own experimental research aimed at identifying and analysing the impacts of significant factors on the thickness of the layer formed. Consequently, the DOE methodology and the statistical processing methodology of experimentally obtained data allowing to derive numerically and statistically correct conclusions was used. Developed mathematical-statistical models, which predict the thickness of the formed protective layer, provided the basis for optimization procedure. The developed models were subjected to another statistical analysis and practical verification in conditions of real production, thus confirming their validity, reliability, or correctness. These models can be used to determine the resulting thickness of the layer formed. They enabled us to derive deposition time, the time of formation of the protective layer and subsequently the so-called critical temperature. The criterion of optimality was set to minimize the deposition time to reduce input costs. We have ultimately contributed to increasing the productivity and efficiency of the process itself while maintaining other required quality indicators. This was ensured thanks to the correct formulation of the optimization problem, specifically the objective function and its constraints (technological conditions of a process). We have programmed all optimization procedures for the solution of the described models with the possibility of implementing real constraints and data using MATLAB system and its special toolboxes. By implementing an optimization procedure for galvanizing and anodic aluminium oxidation, we have determined the optimal combinations of input variables to achieve a minimum deposition time while maintaining the required thickness and quality of the formed layer.

When implementing the technological processes, either machining or surface treatment processes, it is important to ensure that several requirements are considered: both technological (required quality of outputs, such as roughness, layer thickness) and economic (cost minimization, time minimization). Optimization, which is a powerful tool for improving technological processes, has an irreplaceable place in this area. Optimal process conditions of the given technological processes were calculated and consequently it was possible to meet the requirements of the end customers. At the same time, process efficiency improvement regarding the economic aspect was ensured.

References

1. Davim, J. P. (2001). A note on the determination of optimal cutting conditions for surface finish obtained in turning using design of experiments. *Journal of Materials Processing Technology, 116*(2–3), 305–308. https://doi.org/10.1016/S0924-0136(01)01063-9
2. Garg, A., Lam, J. S. I., & Gao, L. (2016). Power consumption and tool life models for the production proces. *Journal of Cleaner Production, 131*, 754–764. Elsevier, Ltd. https://doi.org/10.1016/j.jclepro.2016.04.099.
3. Halická, M., Brunovský, P., & Jurča, P. (2009). *Optimálne riadenie. Viacetapové rozhodovacie procesy v ekonómii a financiách* (200 p.). Epos, Bratislava.
4. Michal, P., Vagaská, A., Fechová, E., Gombár, M., & Kmec J. (2016). The application of predictive model to describe and control technological process. In *Proceedings – 2015 International Conference on Computational Intelligence and Communication Networks, CICN 2015* (art. no. 7546364, pp. 1571–1574). https://doi.org/10.1109/CICN.2015.298.
5. Schrijver, A. (2005) On the history of combinatorial optimization (Till 1960). In K. Aardal et al. (Eds.), *Handbooks in OR & MS* (Vol. 12, 68 p.). Elsevier B. V.

GPSR Compliance

The European Union's (EU) General Product Safety Regulation (GPSR) is a set of rules that requires consumer products to be safe and our obligations to ensure this.

If you have any concerns about our products, you can contact us on ProductSafety@springernature.com

In case Publisher is established outside the EU, the EU authorized representative is:

Springer Nature Customer Service Center GmbH
Europaplatz 3
69115 Heidelberg, Germany

Batch number: 10771034

Printed by Printforce, the Netherlands

explanations can be very effective in repairing trust in cases where the explainer draws on the sheer solidarity of family and kinship. This is particularly understandable as African entrepreneurs are quick to ignore distrust perpetuated by family members. This is because in Africa, trust repair is quickly facilitated by family members who ensure that conflicts are speedily resolved. In such cases, senior family convene hearings to ensure that both parties are provided with the opportunity to give their accounts of the issues (Okrah, 2013; Osei-Hwedie & Rankopo, 2012). However, it is recognised that in some related cases, such cultural norms are not welcomed by other entrepreneurs who tend to disassociate dealing with entrepreneurs who are also family members (Omeihe, 2019).

Apologies in African Entrepreneurship

Although explanations are adequate for trust repair, apologies appear to be more effective as they are perceived to be more sincere. Apologies have a significant impact as defaulting entrepreneurs can personalise their explanations. In turning to research evidence, Omeihe (2019) established that apologies create a strong basis for trust repair among African entrepreneurs. This is because apologies are seen as taking ownership of creating distrust. The most cautious tactic a defaulting entrepreneur can take is to display emotional expressions of regret, while at the same time signalling the intention to act in a more benevolent way. By expressing regret, defaulting entrepreneurs often make personal visits to the victim as a sign of respect. Most strikingly, apologies are most effective when they are timely, credible and sincere (Gillespie & Dietz, 2009; Kim et al., 2004).

Upon closer examination, apologies are more effective when the perpetrator is sincere and assumes responsibility for initiating the violation. Conveying admissions of regret and responsibility, and the desire to reconcile and rebuild the relationship, are found to be important steps in repairing trust after trust violation in African entrepreneurship (Amoako and Lyon, 2014; Omeihe, 2019).

Additionally, apologies for which the violator accepts responsibility for the actions are often in most cases very effective than when the violator

Table 6.2 Tactics and approaches

Repair tactics	*Approaches*
Verbal explanations	Short term
Apologies	Short term
Use of intermediaries	Long term
Use of threats	Long term

this clarification, we proceed now to examine how trust is repaired in African entrepreneurship.

Verbal Explanations in African Entrepreneurship

Verbal explanations appear to be a commonly used tactic for trust repair. Across African societies, entrepreneurs consider verbal explanation as a requisite approach for reconciliation. Earlier work (e.g. Omeihe, 2019) studying the impact of verbal explanations found that explanations were adequate in determining if trust could be repaired. The findings report that a majority of African entrepreneurs request explanations from their defaulting partners in seeking to repair trust relationships. Here, the adequacy of a defaulter's explanation for deception is a major factor for repairing trust. In fact, this determines whether the victim would accept the explanation and, perhaps, indicate the willingness to rebuild trust.

Verbal explanations acknowledge that a violation of agreement occurred, and that the violator has chosen to take responsibility for the violation. However, the nature of explanations also matters. For instance, explanations which are detailed and specific of the violation are considered more productive than less specific ones. As far as Africa is concerned, many entrepreneurs consider verbal explanations as important for clearing the air of distrust, especially in situations where there was no preconceived intention to violate trust. Verbal explanations are also seen to provide swift resolutions to grievances. Simply repairing damaged trust, however, is dependent on the adequacy of the defaulting entrepreneur's explanation. A more lasting solution would demand that the defaulting entrepreneur assumes ownership and responsibility for the damaged trust.

Complementing verbal explanations, socio-cultural norms play an important role in trust repair efforts. In the African context, verbal